JUST DO THIS:
BACK UP!

JUST DO THIS:
BACK UP!

A Step-by-Step Plan To Back Up Your Computer

Version 1.0

by

Leo A. Notenboom

An Ask Leo!® book

https://askleo.com

JUST DO THIS: BACK UP!
A Step by Step Plan To Backup Your Computer
Leo A. Notenboom

ISBN: 978-1-937018-28-3 (paperback)

Contents

I believe personal technology is key to humanity's future.

It has an amazing potential to empower individuals.

But it can also frustrate and intimidate.

I want to make technology work for you.

I want to replace that *frustration* and *intimidation*

with the *amazement* and *wonder* that I feel every day.

I want it to be a *resource* rather than a *roadblock*;

a *valuable tool,* instead of a source of *irritation*.

I want personal technology to empower you, so you can be a part of that amazing future.

I just want it to work, for you.

That's why Ask Leo! exists.

Leo A. Notenboom

https://askleo.com

Your purchase of this book includes several additional bonuses:

- Videos that walk you through the steps outlined in several of the chapters in this book.
- All available digital formats of the book as direct downloads, so that regardless of which version you purchase, you can enjoy this book on the digital device of your choice.
- Updates, errata, and prioritized Q&A.
- Any additional bonuses that might be made available.

You'll find the information that you need to register in a chapter called "Register Your Book!" near the end of the book. Once you register you'll have access to a website area specifically for this book that will have all available bonuses and updates.

Backing up doesn't have to be hard.

The concept - just making copies of everything important - is about as simple as it can be. Put another way (and as something I say on Ask Leo! All the time):

If there's only one copy, it's not backed up.

Unfortunately, people can get confused at a number of steps along the way:

- Figuring out what, if anything, to buy.
- Understanding what's "important" and in need of backing up.
- Figuring out how to make backups happen automatically.
- Worrying about the security of backups.
- Backing up things where the very concept of "just make a copy" isn't at all obvious.

Much of the confusion comes from the fact that there are *so many options* and ways to accomplish all this. Ask a dozen different people what the "best" way to back up is, and you'll get a dozen different answers if not more.

There are hundreds of ways to approach the problem of backing up. So many so that… well, one of my readers said it best:

I was immediately so overwhelmed that I gave up.

I understand. It's confusing.

That's why I decided to focus on a "Just Do This" approach to backing up.

Just Do This

I'm going to outline for you one arrangement for backing up your desktop or laptop PC that just works. Follow these instructions, and you'll be backed up.

I'll make the hard decisions for you. Together we will:

- Get an external hard drive.
- Make an image backup.
- Schedule automatic image backups.
- Install and configure Dropbox for near-continuous backup of your work in progress.
- Install and configure BoxCryptor Classic to protect your valuable data in the cloud.
- Install and configure Dropbox on your smartphone, for immediate backup of the photos you take.
- Install Thunderbird and configure it to back up your on-line email account.
- Make a list of a few manual maintenance tasks choose to do on a regular basis.

Don't worry:- I'll give you guidance if you want to do things slightly differently. Even if you do follow this plan to the letter, you'll be able to make changes later, as your needs, and your own comfort level, changes as well.

Backups are important. So important that if you do nothing else…

Just Do This.

1

Purchase an External Drive

My first recommendation is that you purchase an external hard drive for your backups. Backing up to an external drive is the most important step in getting your overall backup strategy in place.

The inevitable question is just which external drive to get.

The problem, of course, is that the answer keeps changing. Technology evolves, and as a result, so does my recommendation.

Let me give you a few guidelines, and then a few current (as of this writing) examples.

External drive capacity

It's pretty easy to say you can never have too much capacity. Bigger is better.

Of course, more capacity also means spending more money.

The rule of thumb I like to use is this: get an external drive that is at least three times as large as the amount of data you expect to back up.

For example, one of my Windows 8 machines has a roughly 100GB (gigabyte) drive, of which around 65GB are used.

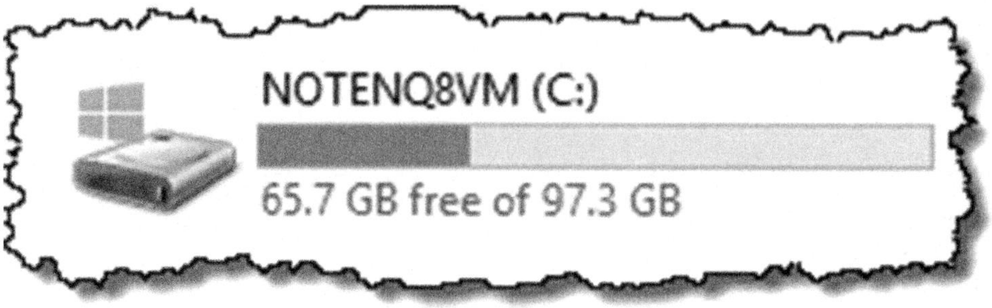

Thus I would recommend a capacity of at least three times 65GB, or around 200GB. That would be enough to hold two complete and compressed full-image backups, along with overhead information such as recovery partitions, and a healthy collection of incremental backups as well.

As I said, that's a *bare minimum*, and there are certainly situations where it could end up not being enough, depending on how you configure your backups. If you want extra safety and breathing room, double or even triple my recommendation.

In most cases, the good news here is that your back-up requirements - even after tripling my recommendation - will likely be smaller than the average external hard-drive size currently available.

External drive technology

There are, naturally, several different types of external drives. Connection methods and even power options vary. Some choices are easy; some depend at least a little on your personal setup.

USB3

There's simply no reason not to ensure that your new external drive comes with a USB3 interface, even if your computer doesn't support it.

If your computer does support USB3, backups will be faster - potentially much faster - than the previous USB2 standard. If your computer doesn't support USB3, that's okay; USB3 is backwards compatible, and will simply operate at the slower USB2 speed.

Someday, when you get a new computer that will, in all likelihood, have a USB3 interface, you'll have your external drive all ready to take advantage of it.

SSD

In my opinion, SSDs, or Solid State Disks (which use high-quality flash memory instead of rotating magnetic platters), aren't really worth the money for back-up purposes. I say this for two reasons, and probably not the reasons you expect.

- Expense: even though they are coming down in price, SSDs are more expensive than traditional hard disks.
- Longevity: the lifespan of magnetic media is pretty well understood. Once written, it stays written for a long time, and even in the worst circumstances, data can typically be recovered from it (albeit at a price, depending on many factors). I'm just not comfortable with SSDs for really long-term storage yet.

Even though I recommend getting USB3 for the speed, there's no real harm or additional cost involved in getting it. The truth is, when it comes to backups, as long as they complete in a reasonable amount of time, speed isn't all that important. Thus, the primary feature of SSDs - speed - is unwarranted. In fact, it's doubly unwarranted, since the speed advantage of an SSD is mostly about *reading* data, and backups are all about *writing* to the drive.

What I didn't mention was 'wearing out'. Flash memory does wear out the more you write to it, but SSDs have a longer useable lifespan than their cheaper thumb-drive counterparts, and backups actually don't write as much to the drive as you might think. Yes, they write a copy of everything (or everything that's changed) each time you back up, but that's nothing compared to the constant use that SSDs are under - and are designed to handle - as the primary disk drive of a computer.

All that being said, there's nothing wrong with using an SSD as a backup drive, as long as it is a true HD-replacement quality drive, and not a cheap thumbdrive. I just wouldn't recommend spending the money on it.

Power

There are two approaches to powering an external drive.

- No additional connection: the drive is powered entirely by the USB interface.
- An separate external power supply.

The choice is yours, but when making your decision, consider that USB-powered drives are typically small, portable, and slower. Drives with external power supplies are typically physically larger as well as having more capacity, and are able to transfer data more quickly.

I use both, but as it turns out, both of my back-up drives for my main machines are USB-powered. The one with my laptop is an obvious choice: it's small, portable, and easy to travel with. It just turned out that the drive I use for my desktop is also a USB-powered drive - as I said, I'm not concerned with speed.

Other characteristics

Physical size: if you care, you can narrow your selection based on physical drive size - for example, a 2.5 inch drive versus a 3.5 - but in the long run, that choice will be made for you based on the choices you made above. USB-powered portable drives are generally small, and the drive with external power (and possibly higher capacity) will be physically larger as well.

Rotation speed: I never pay attention to this for external drives, particularly back-up drives. The USB interface speed will likely be the limiting factor, so focus instead on getting USB3, if you can. If you're on USB2, the speed of a faster drive is mostly wasted.

Specifics: what external drive to get

As I said, drives are changing constantly. With that in mind, here are a couple of drives that I can recommend today - where "today" is January 2015[1].

1 You can find updates online, of course, at https://askleo.com/external-drive-get/

USB-powered: Seagate 1TB USB3 Portable External Hard Drive[2].

2 http://go.askleo.com/ama/B008R7FC74b

I have three of these. It's the drive that sits next to my laptop, my desktop, and even my wife's laptop as well, so you know I trust it. I recently relied on the one I use to back up the laptop when I upgraded the laptop's internal drive, and it worked flawlessly.

All of mine are the 1TB (terabyte - that's a thousand gigabytes) version, but a 2TB is available as well, and if I were to buy it again, that's probably what I would get today.

Externally powered: Seagate 2TB USB3 3.5-Inch Desktop External Hard Drive[3].

Even though I don't use it for backup, I use one of these as a day-to-day work disk for my desktop machine. The USB3 interface means it's fast - fast enough for my day-to-day work. Even though that speed might not be required for backing up (which actually

3 http://go.askleo.com/ama/B00834SJS0

applies to the USB-powered drive above, as well), I have no hesitation recommending it as a back-up drive.

Mine happens to be the 2TB version, but checking on it as I write, I see that there are 3TB, 4TB, and even a 5TB version available as well, and an 8TB version was just announced. They'd be hard for me to resist. Like I said, you can't have too much disk space.

Opinions are easy to come by

Hard disks are a very difficult category of product to recommend. The problem is that the industry is cyclic: a great hard-drive manufacturer five years ago might be horrible today.

Unfortunately, that's true for all the major hard-drive manufacturers:their quality appears to arrive in waves. Fortunately, there's usually at least one cresting while another is at its low point.

As you can see from my recommendations above, I've been pretty happy with my Seagate drives for some time. Naturally, that can change. But it also means that others may feel differently, either based on their own out-of-date experience, or because the market has changed.

So, when you're shopping for an external hard drive, pay attention not only to the reviews you might read, but also to the dates of those reviews.

Just Do This:

If you don't have one already, get an external hard drive for backing up. Either of the two I've mentioned above, or something similar, will do.

2

Make a Backup Image

The very first thing you want to do with your new external hard drive is to make an *image* of your computer's hard drives. An image is nothing more than a copy of absolutely everything.

In fact, making an image backup is what I recommend you do first when you get a new machine as well. A backup image can substitute quite nicely for missing installation media should you ever need it.

More importantly, creating backup images will be the basis for our backup strategy. First we'll make one manually, and then in the next chapter we'll look at making it automatic.

Get the software

I'm going to use the free version of Macrium Reflect to make my initial system image. It's perfect for this purpose. There are other tools, of course, and you're certainly welcome to give them a try. The basic concepts will be the same.

Start by downloading Macrium Free from here:

http://go.askleo.com/macriumfree

that'll take you to the page on the Macrium site that has the free version of the tool.

Take notice: many software manufacturers offer both free and paid versions. Make sure to select the one marked "FREE", not "Free Trial". A trial implies that the version you're getting isn't free at all, and will eventually require payment. Macrium's free product is truly free.

As I write this, the Macrium site actually hosts the download of the free version on c|net's download.com. I strongly recommend that you always start at the manufacturer's site - macrium.com in this case - and click on the download link *there* that takes you to download.com.

That way, you know you're getting the officially provided version, and not a fake, malware-laden knock-off.

When you're on the download site, take care not to mistakenly click on ads that look like the download.

Ads are constantly trying to "look like" free downloads when in fact they are nothing more than ads.

The download is actually Macrium's small "Download Agent". Run that.

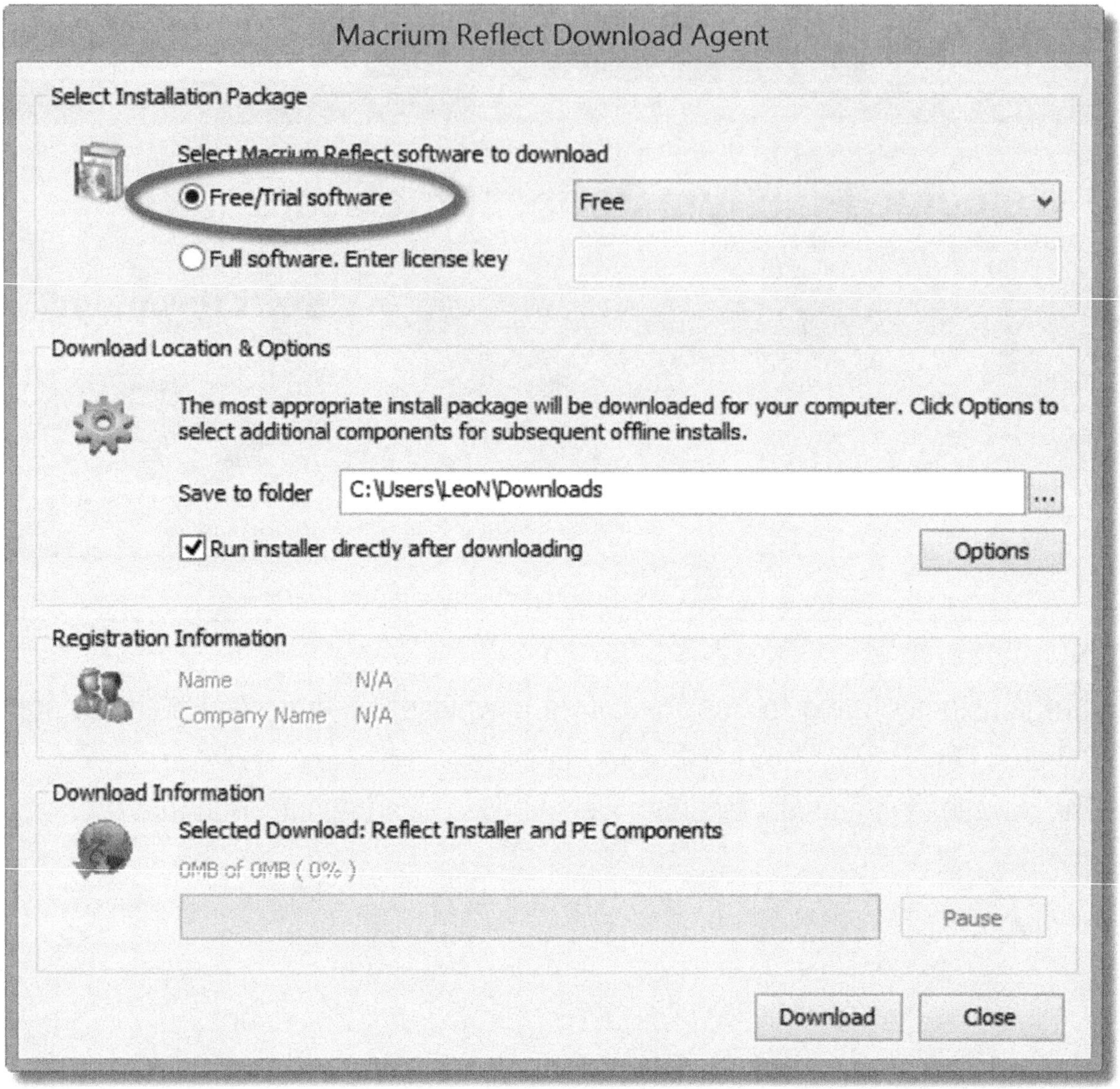

Make sure to select "Free/Trial software", leaving "Free" in the selection to the right, and click on Download to download Macrium Reflect free. It may be a large download (~450 megabytes).

The default options are fine for Macrium, though you don't need to register unless you want to.

Disconnect new machines from the network

This seems counter-intuitive, but if you have a new machine, I want you to do this all as early as possible, because you may not have all of your security and other customizations in place just yet. While it's very likely that the default configuration of your new machine makes it safe to download Macrium Reflect, I don't want to take any further risks.

Once you've completed the new machine image, then you can pick up after the "Make a system image backup" step in my article: The First Eight Things To Do With Your New Computer.[4]

For now, disconnect from the network.

(If you're creating an image of a machine that is fully configured and secure, you don't need to disconnect.)

Attach your external hard drive

In the previous chapter I had you purchase an external hard drive for your backups. I strongly recommend using an external hard drive to hold the image we're about to create, because a system image is really too large to back up using CDs or DVDs.

In a pinch, a USB flash drive can do, but in that case, I recommend that you copy that backup to another location for safekeeping.

Regardless of the location, make sure that it has enough free space to hold the contents of the disk you're backing up. Yes, the backup will be compressed, but by ensuring that there is more than enough space, you'll also have room for the partitions and overhead information that you might not easily be able to include in your calculations.

Note: While Reflect can back up across a network, if you followed the previous step for a new machine, you're no longer connected to a network. For an existing machine that's already properly secured and still connected, backing up across the network works as well.

4 https://askleo.com/4689

Make the image backup

Setup will have left an icon for Macrium Reflect on your desktop.

Double click that to start Reflect.

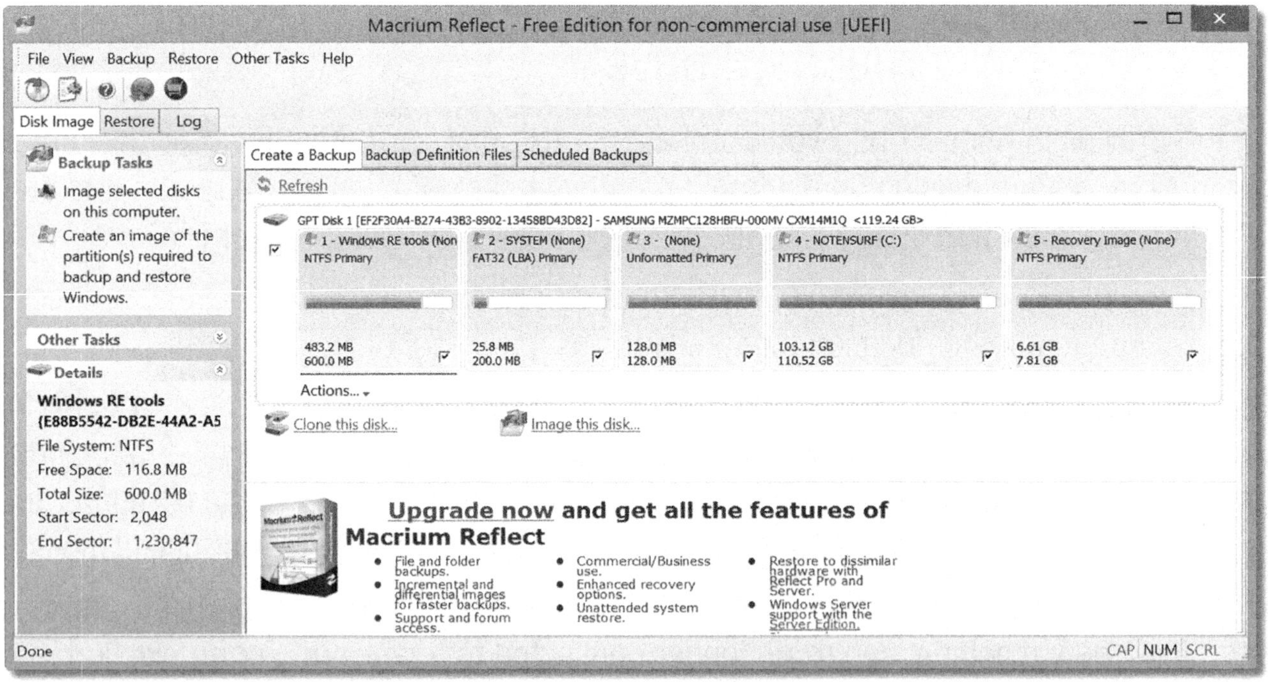

Reflect opens to a display of the partitions found on your computer[5]. You may have many partitions, or just a couple.

5 The example machine here is my original Microsoft Surface Pro.

Fortunately, you don't actually need to know or care what partitions you have. Click on "Create an image of the partition(s) required to backup and restore Windows", which is on the left side of the Reflect Window.

This will bring up the partition selection dialog, with all the partitions that you need to back up automatically selected.

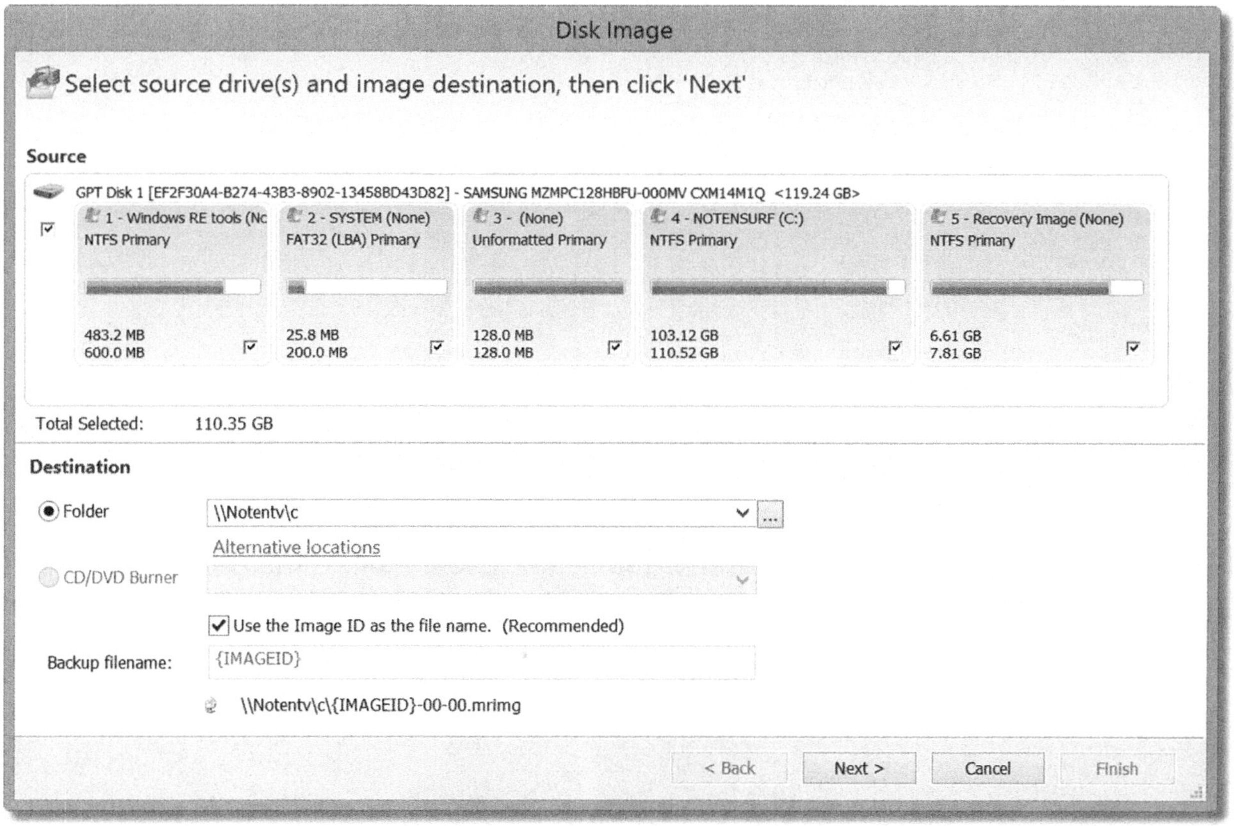

Typically all the partitions on your hard drive (if you have only one) will be selected.

Under Destination, select "Folder", and type in the drive and folder where you want to place the backup image. Alternately, use the "three dots" (ellipsis) button to open a destination-selection dialog.

If you're using an external hard drive or USB drive, the destination may be as simple as "D:\", where "D" would be replaced by the drive letter assigned to that device.

Click on **Next**. This will give you a summary dialog of what's about to happen.

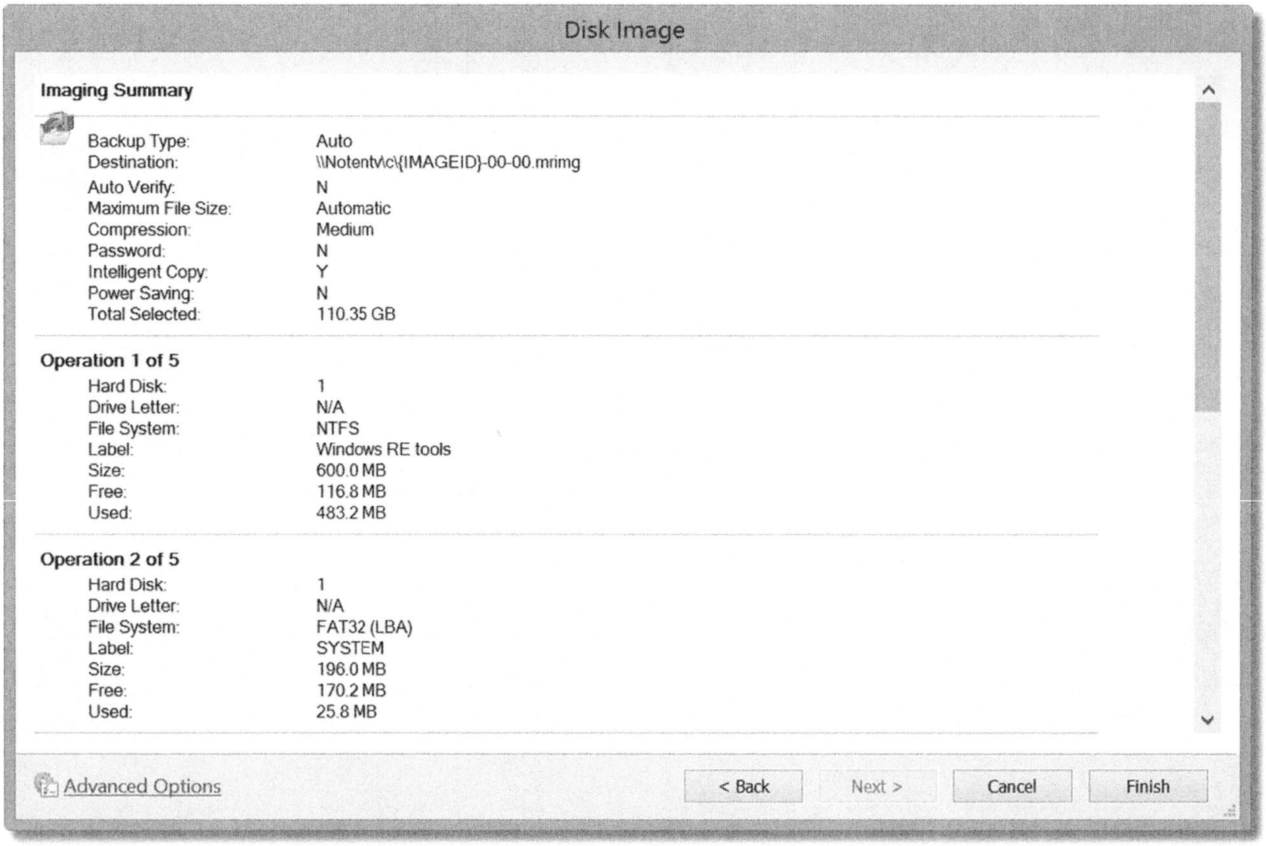

If you're curious, you can look around at the list, or even explore the Advanced Options, but nothing need be changed. Click on **Finish**.

You've just created a "backup definition". Now you can choose to save it to a file for use later.

For our purposes, nothing needs to be saved here, so you can uncheck "Save this backup as an XML Backup Definition File". Leave "Run this backup now" checked, however, and click **OK**.

The backup begins.

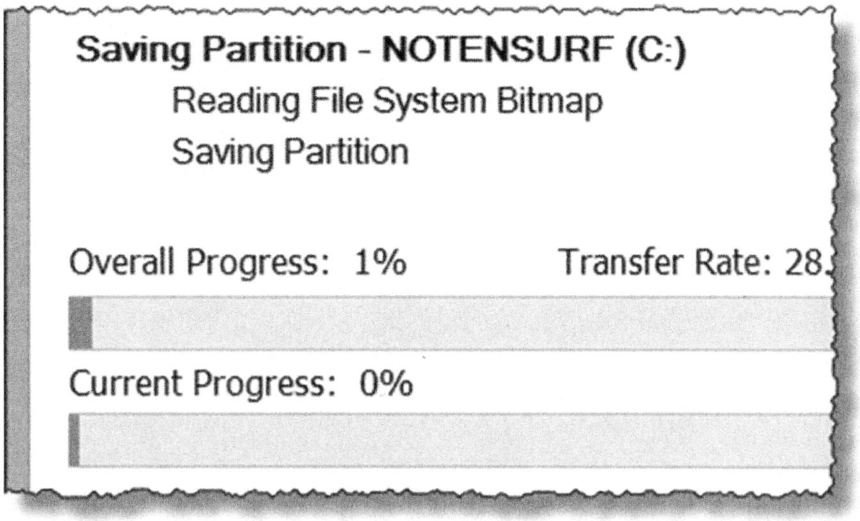

How long the backup takes will depend on many things, including how much data there is to back up, and how quickly the data can be written to the destination.

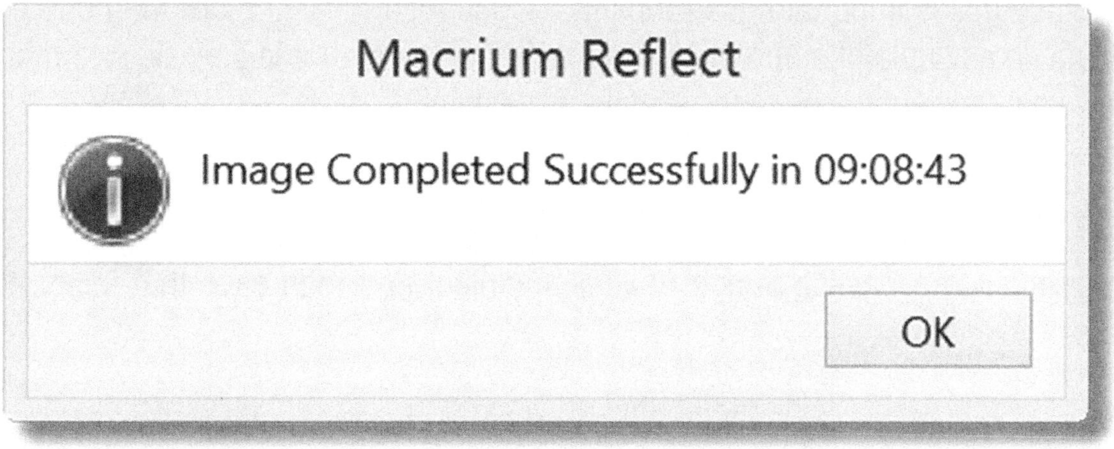

Your backup image is complete.

Save your new machine image

The backup image you've just created is important: you'll want to save it somewhere safe. If, for example, you've created it immediately after getting a new machine, it takes the place of your original installation media. Should you need to reset your machine to "factory settings", you'll restore this image to your machine.

Reflect images are stored as ".mrimg" - for Macrium Reflect IMaGe - files. Reflect assigns the file a name made of seemingly random string of letters and numbers. For normal periodic backups that's probably just fine, but for something like our new machine image, you may want to rename it to something more meaningful. Just remember that it must end in ".mrimg".

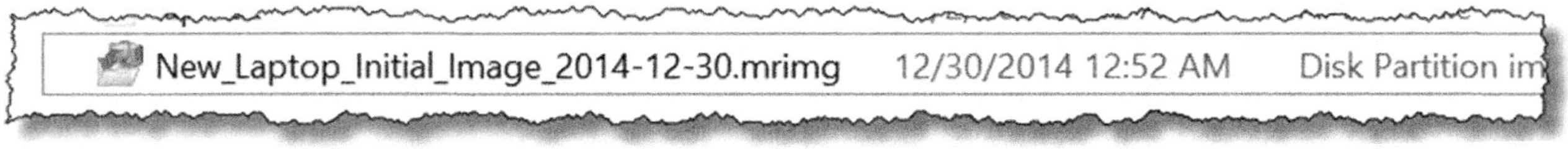

Again, since this is important, it's probably a good idea to save more than one copy of this image in a couple of different places that you'll remember in a week, or a month, or in a few years, when you need to reset your machine.

Just Do This:

Begin by making a backup image of your machine, and then save that image somewhere safe.

3

Schedule Regular Backup Images

Next to simply making an image backup, scheduling those backups to happen automatically is one of the topics most people find confusing.

Depending on the backup software that you're using, it's typically not hard at all.

Now that we've installed Macrium Reflect Free and made an image, I'll show you by walking through the steps of scheduling a monthly full backup.

Defining the backup

We just walked through the steps to create a one-off image backup.

Now we're going to build on that, and set Reflect to make a new system-image backup automatically once a month.

In the previous chapter, we:

- Installed Macrium Reflect Free
- Defined a backup of the primary disk in our machine
- Defined a location to place the backup image file
- Created that backup image

The process for defining the monthly backup is the same, but I'll make one small change.

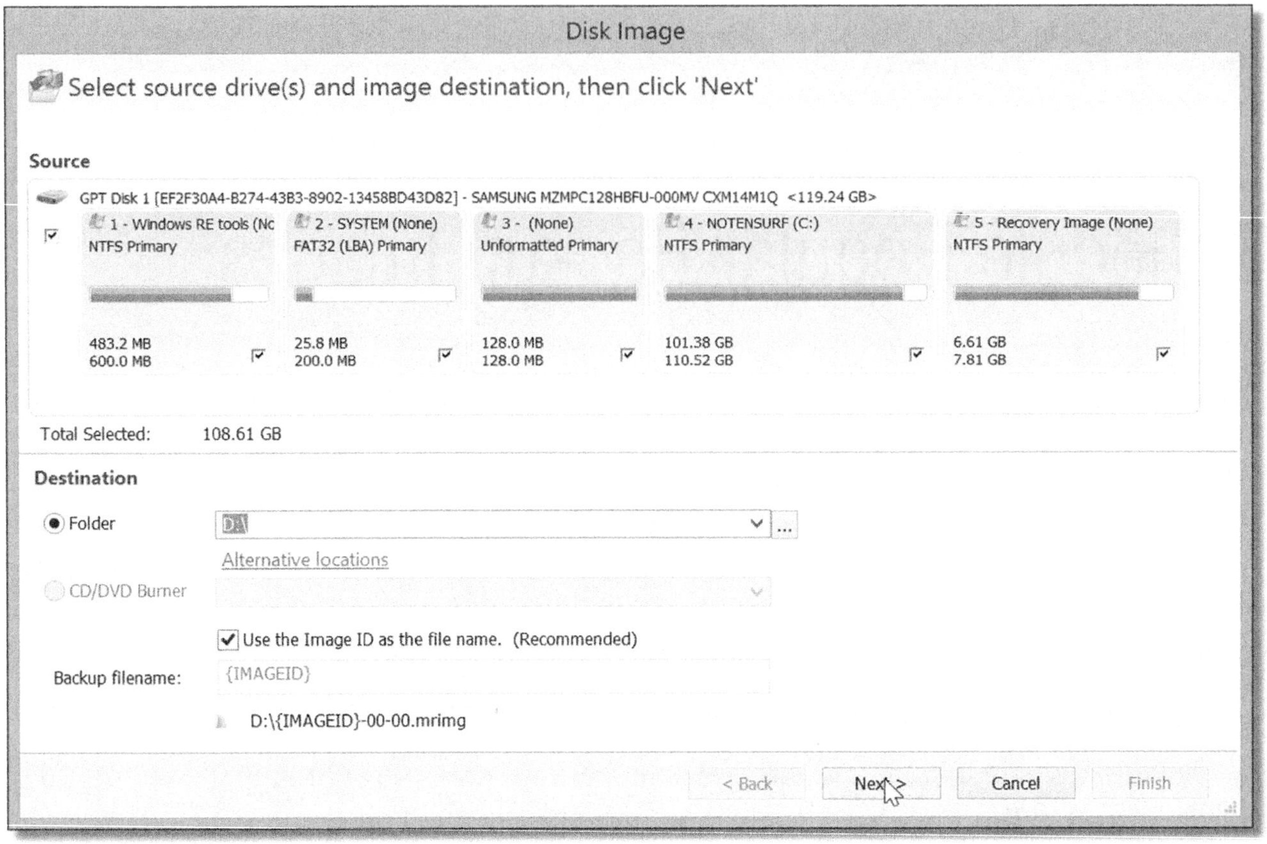

"Use the image ID as the file name" remains checked. To avoid overwriting existing backup images, I let Reflect simply use the Image ID, which is different for each image created, as the filename.

The other change comes when it's time to either run the backup or save the backup definition file.

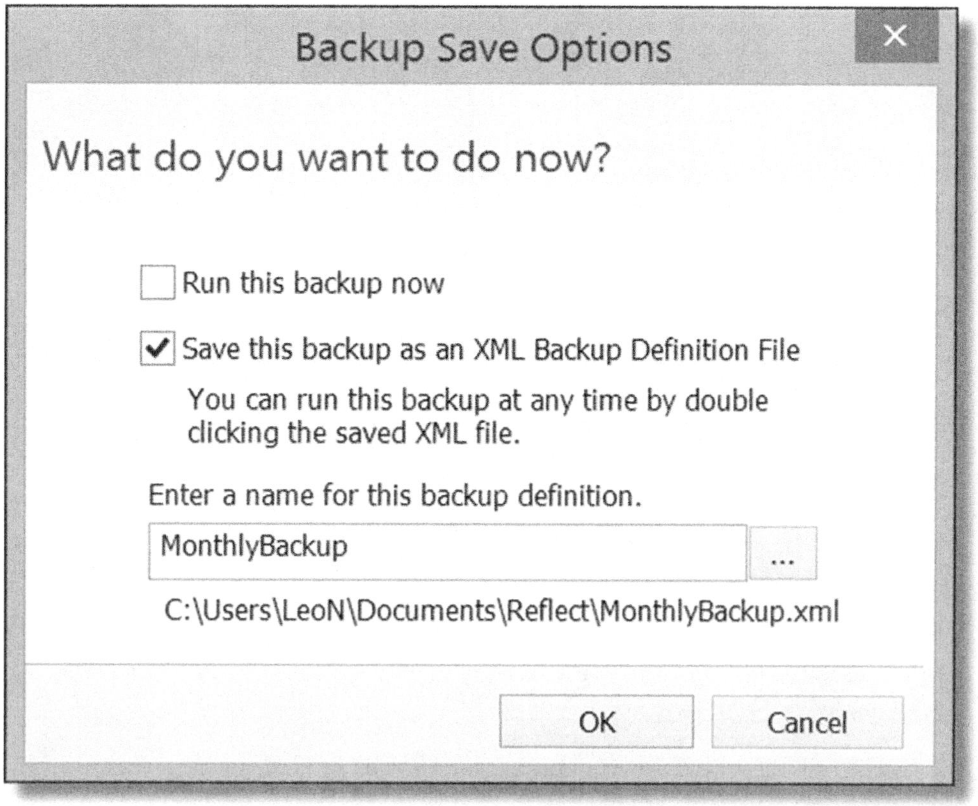

I've unselected "Run this backup now", and instead selected "Save this backup as an XML Backup Definition File". This causes all the decisions made to this point - what to backup, where to place the backup image, and so on - to be saved. I've also elected to give that backup definition file a name: "MonthlyBackup".

Schedule automatic backups

After saving the XML backup definition, Reflect should bring you to the Backup Definition Files tab. If not, click on that tab to see a list of all the backup definition files you've created.

Right click on the one we just created, **MonthlyBackup**:

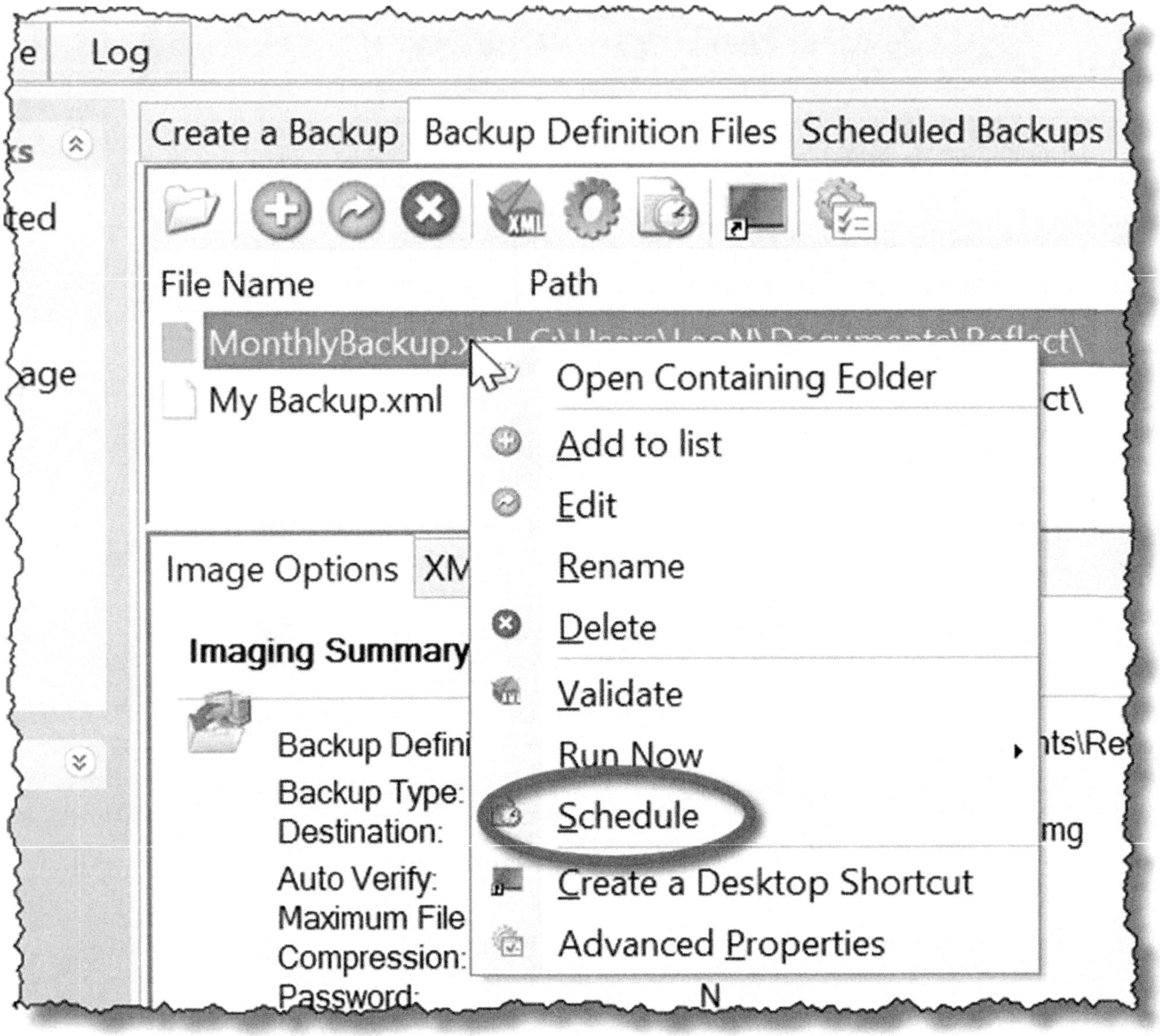

Click on **Schedule** in the resulting menu.

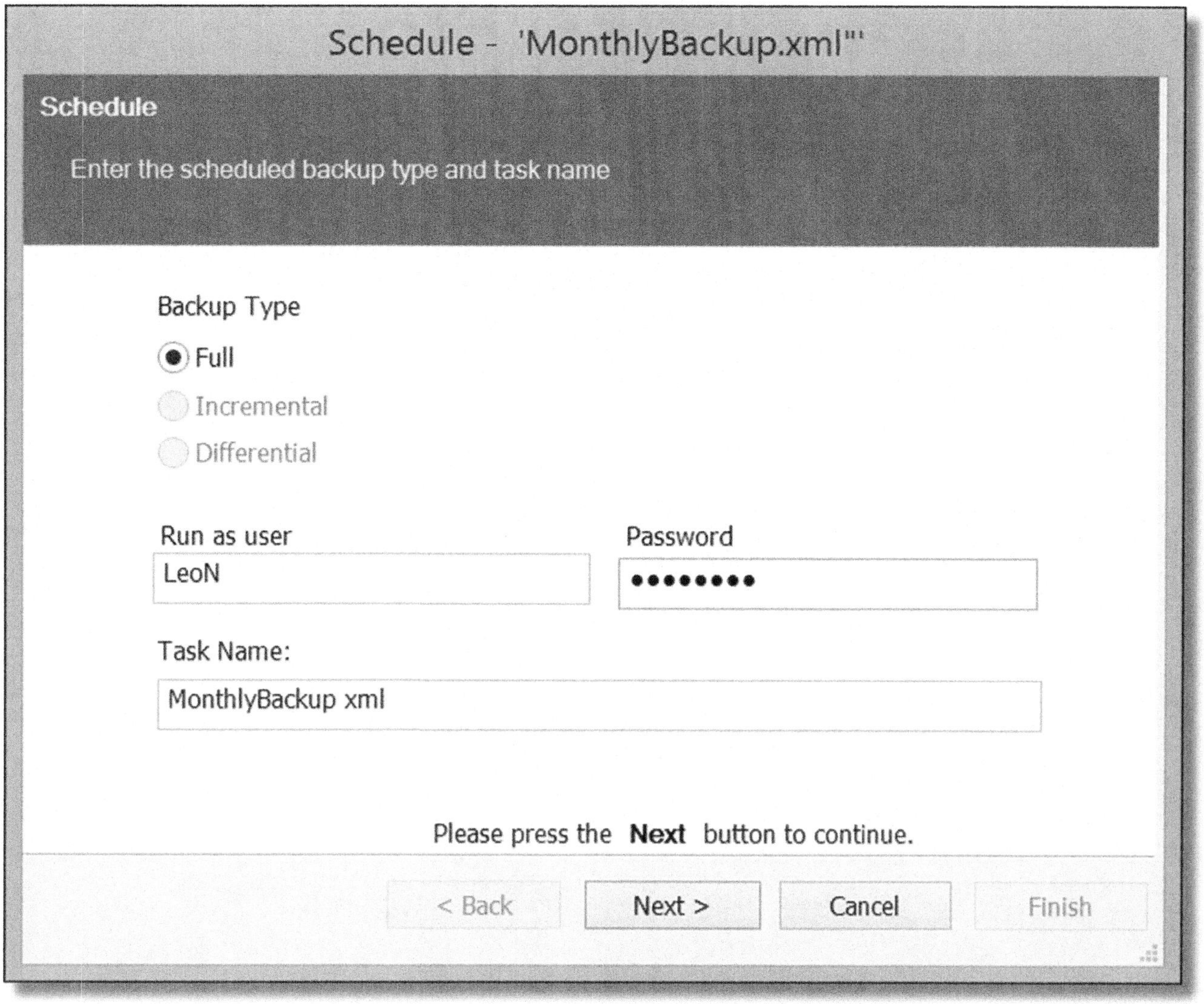

The only choice of Backup Type in the free version of Reflect is “Full”.

Enter your Windows user ID and password - the same user ID and password you use to log in to Windows itself. Reflect needs this so that it can run your backups whether or not you’re logged in.

The Task Name is up to you. I tend to leave it as the name of the xml file I originally created.

Click **Next**. In the next dialog box, we select how often the backup should be run:

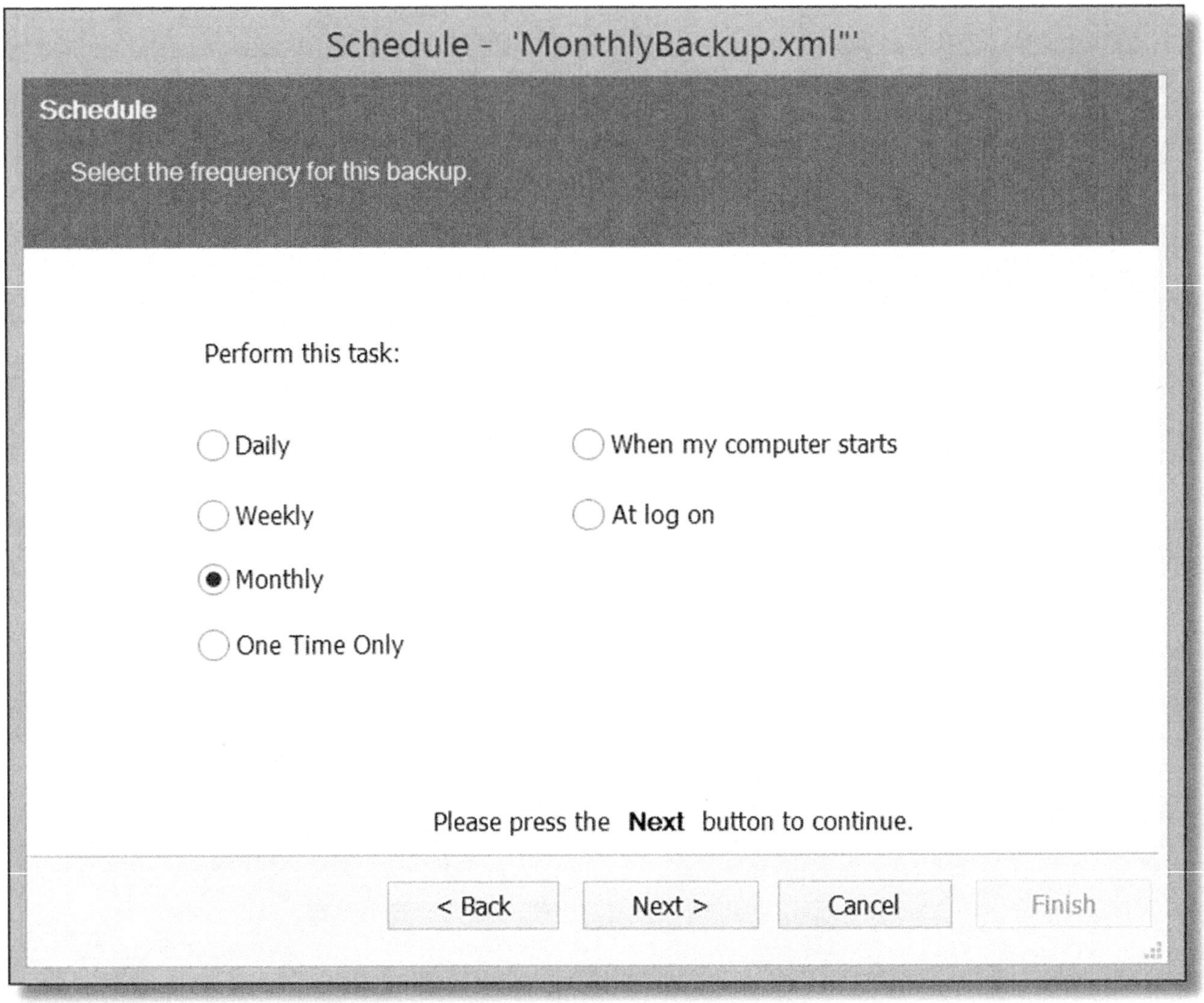

Select Monthly and click **Next** to set up the specifics.

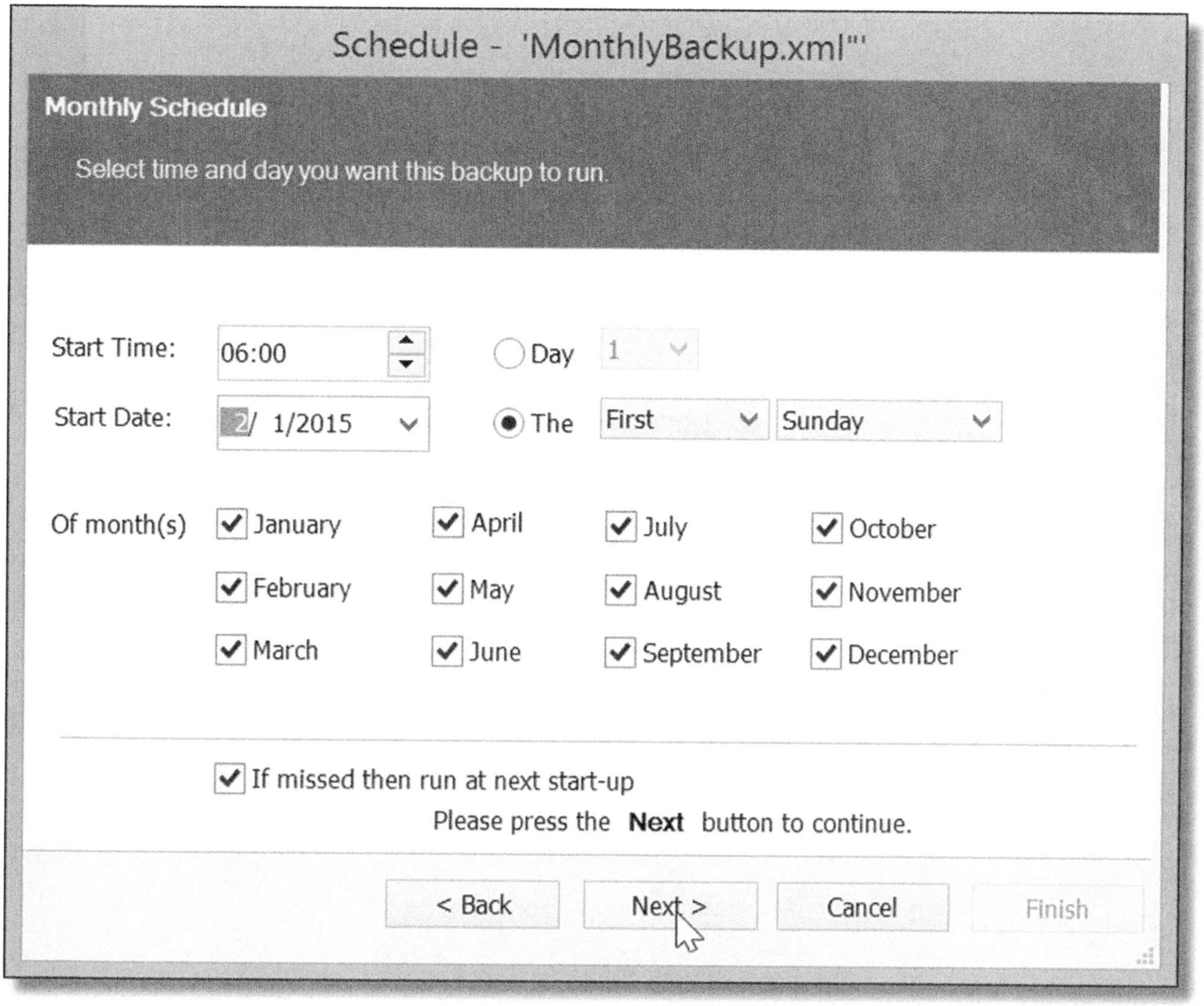

For my example, I've set up the monthly backup to happen at 6:00 a.m. on the first Sunday of every month.

One important option that I recommend you select is "If missed then run at next start-up". If for some reason your machine is unable to run your backup at its scheduled time - perhaps the machine is turned off - then it'll run automatically the next time you log in.

Click **Next** and you'll be presented a summary of the scheduled operation:

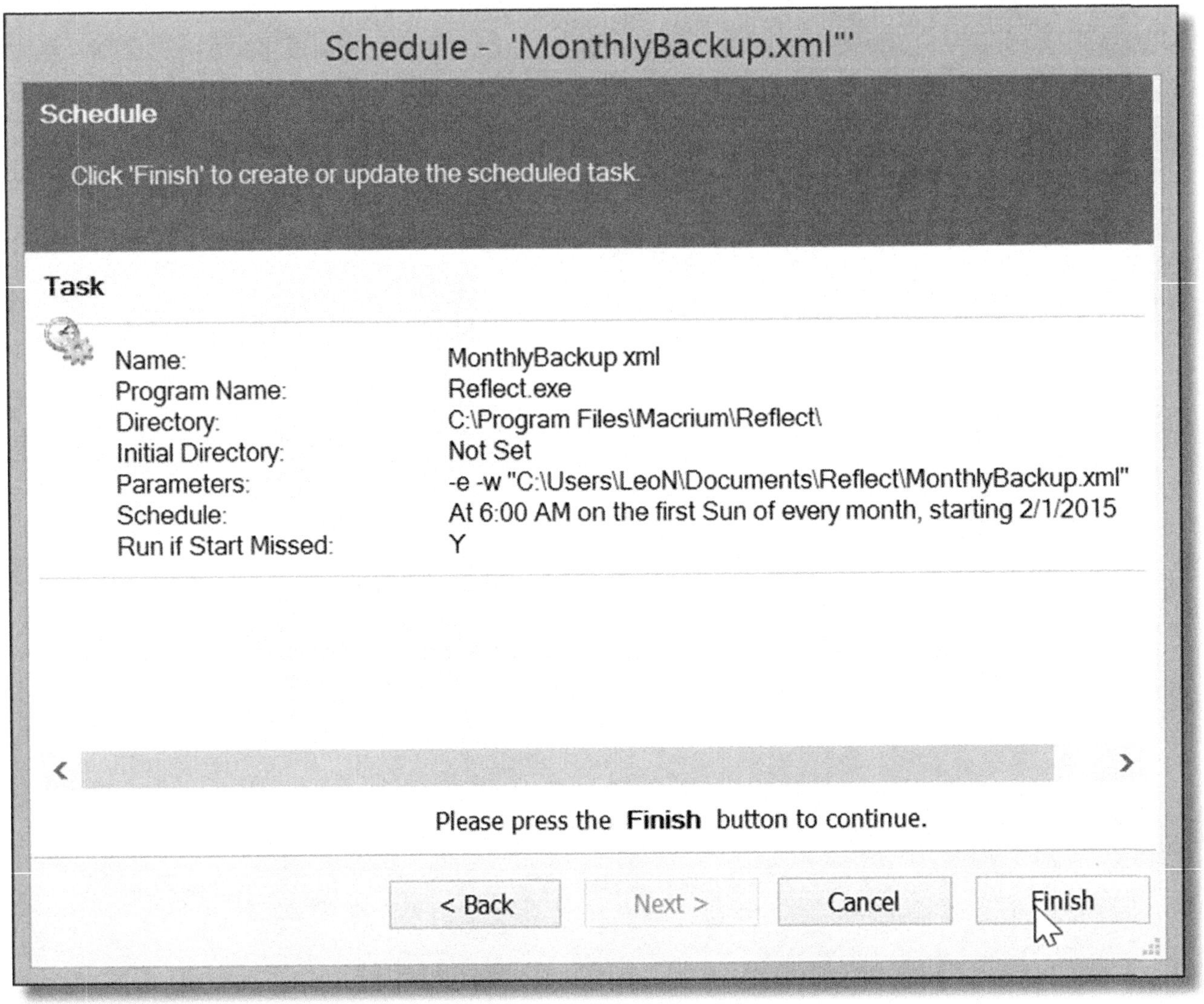

Click on **Finish**, and ... you're done!

Your backup is now scheduled and will happen once a month, automatically, without any thought on your part. You can confirm by examining the contents of your backup drive after the next scheduled backup and making sure that a new image file has been created.

Automatic backups and disk space

At this point, you may be asking yourself if these automatic backups are simply going to accumulate, taking up more and more disk space.

The answer is yes.

In my example configuration, each month Reflect will create a new backup image, and add it to the collection of images that have been made to date.

Eventually, you're going to run out of disk space on your backup drive.

There are two solutions:

- Periodically delete or move older backup images to free up space for newer images.
- Use the paid version of Reflect, which includes two features that can help:
- **Incremental backups**, which can dramatically reduce the amount of space that backups take, because they back up only those things that have changed since the previous backup. Since many files don't change once installed, this provides a significant reduction in the rate at which disk space is used up.
- **Disk space management**, which can automatically delete older backups based on age or quantity. For example, I use this to keep only the two most recent backup sets, deleting the oldest automatically after the third backup is created.

Naturally, if you're using another backup program, disk-space management solutions will differ. It's an important issue to be aware of, however, so that your most important (newest) backups aren't blocked by a simple lack of disk space.

Is monthly enough?

Monthly backups are a bare minimum. Combined with the technique I'll outline in the next chapter, it represents a good base for many, if not most computer users.

But I'll admit that I've artificially limited my example to monthly for one simple reason: we're using the free version of Reflect, which only does full backups. Monthly represents a tradeoff between backup frequency and the disk space issue we just discussed. You can create image backups more frequently if you like, but you'll use up disk space more quickly, and have to manage it more often, on your external backup drive.

My preferred alternative is to use the paid version of Reflect, and use what are called "incremental" backups. These backups only include the files that have changed since the previous backup, and thus consume disk space at a slower rate. This makes a combination of monthly full backups and daily incremental backups a more complete solution.

However, it does require purchasing Macrium Reflect (Standard or Pro) for the ability to perform incremental backups. With that purchase you'll also get the ability to manage disk space automatically.[6]

Just Do This:

Schedule regular automated backups images of your machine, and keep an eye on the disk space used on your external drive.

You could stop here, but…

You have an image backup of your computer, and it's being created regularly on an automated schedule.

That's *fantastic*. That already puts you ahead of at least half of the typical computer users out there.

But we can do better. You're still at risk for losing work in progress between backups, as well as other important files. The next few chapters will take care of that.

---

6 Macrium Reflect can be purchased directly from http://macrium.com. You can also purchase my *book* detailing how to set up features like incremental backups and disk space management here: http://go.askleo.com/savedmr . Use coupon code JUSTDOTHIS for 30% off the cover of the PDF version.

4

Use Dropbox for Near-continuous Document Backup

We've successfully set up monthly image backups of our machine. That's a *huge* step in the right direction.

But what about the files we work on frequently - the data that changes between those monthly backups? The data that sometimes changes multiple times a day?

That's where tools like Dropbox come in.

Dropbox and Dropbox-like programs

I'm going to use Dropbox as an example throughout this chapter, as it's an ubiquitous and popular utility. If you prefer, there are alternatives, including SugarSync, Box.com, Spideroak, Google Drive, Microsoft OneDrive, and more.

The key feature that these utilities provide is automatic syncing of documents between multiple computers. If you create or save a document on machine "A", it is automatically replicated or updated on machine "B" (and "C" and "D" and however many you are using with the utility).

However, you don't need two (or more) machines for these utilities to have value. Even if you only work on one machine, the value comes from a by-product of the data synchronization. What looks to you and me like a direct copy from machine "A" to machine "B" is in fact[7]:

- A copy from machine "A" to the service's cloud storage.
- A copy from the service's cloud storage to machine "B".

Between those two steps, your material is backed up to the cloud. Even if there is no machine "B", you still benefit, because every time you save a document, it's automatically copied to or updated on the service's cloud storage.

What if there is no machine "B"? Then every time you save a document on machine "A", it's automatically copied or updated on the service's cloud storage.

That sounds exactly like a backup to me. Every time you save the document it's backed up.

---

7 I'm aware that on local networks this functionality is often optimized as a direct copy from machine "A" to machine "B", in addition to the copy from "A" to the cloud. That, and many other optimizations, are beside the point of our current discussion.

Installing Dropbox

Go to dropbox.com and sign up for an account.

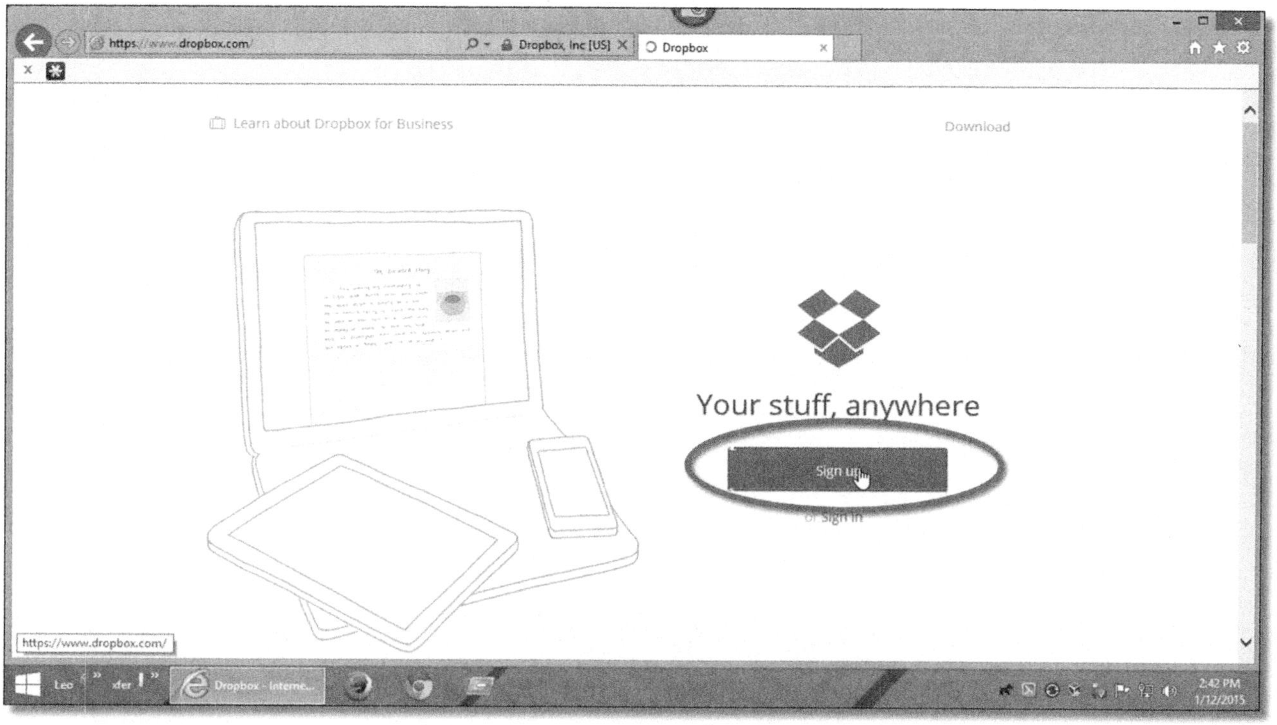

Fill in your name and email and choose a strong password.

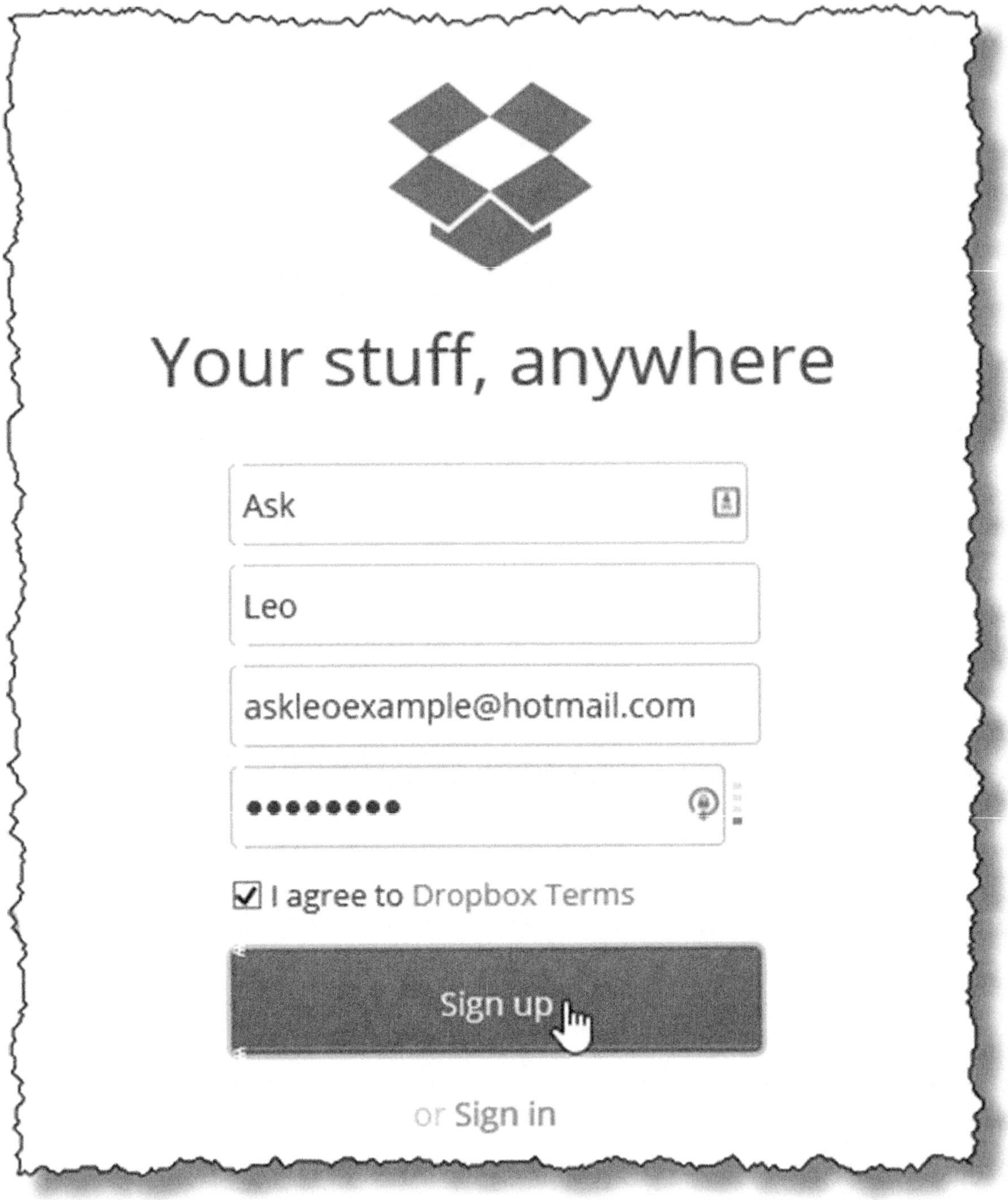

Naturally, you'll have to agree to Dropbox's terms of service. Clicking **Sign up** will take you to a page to select which Dropbox plan you want:

The free two-gigabyte plan is a fine place to start[8].

Once you click **Continue**, the Dropbox downloader will download and install the program.

8 There are ways you can increase your free plan storage to nearly 20 gigabytes, mostly using their referral program. The amount of space included with the free plan is also one of the ways that competitors differentiate themselves from Dropbox.

You can investigate the settings as you like, but the default installation is actually all you need. The important aspects of the newly installed Dropbox are:

- It will automatically start with Windows.
- A Dropbox folder will be created in C:\Users\<username>\Dropbox, where "<username>" is replaced with your Windows username.
- The dropbox icon will appear in the taskbar notification area.
- Dropbox icon in the taskbar notification area

The Dropbox folder

The folder that the Dropbox installer created in C:\Users\<username>\Dropbox is where all the magic happens.

- Anything you place in this folder is automatically copied to the Dropbox servers.
- Anything you update in this folder is automatically updated on the Dropbox servers.

This includes any and all sub-folders that you create.

If you have more than one machine using Dropbox with this same account, then any files placed into this folder or updated on other machines are downloaded (synchronized) to this one.

Note that this is only true for documents within the Dropbox folder.

If you're using Dropbox for automatic behind-the-scenes backup, we need to make one other change.

Changing the default document folder

I'm going to use Microsoft Word for this example, but the concept is very simple: change the default document folder in whatever application(s) you use regularly to be a folder within the Dropbox folder.

In Word 2013, under **File**, **Options** click on the **Save** sub-section:

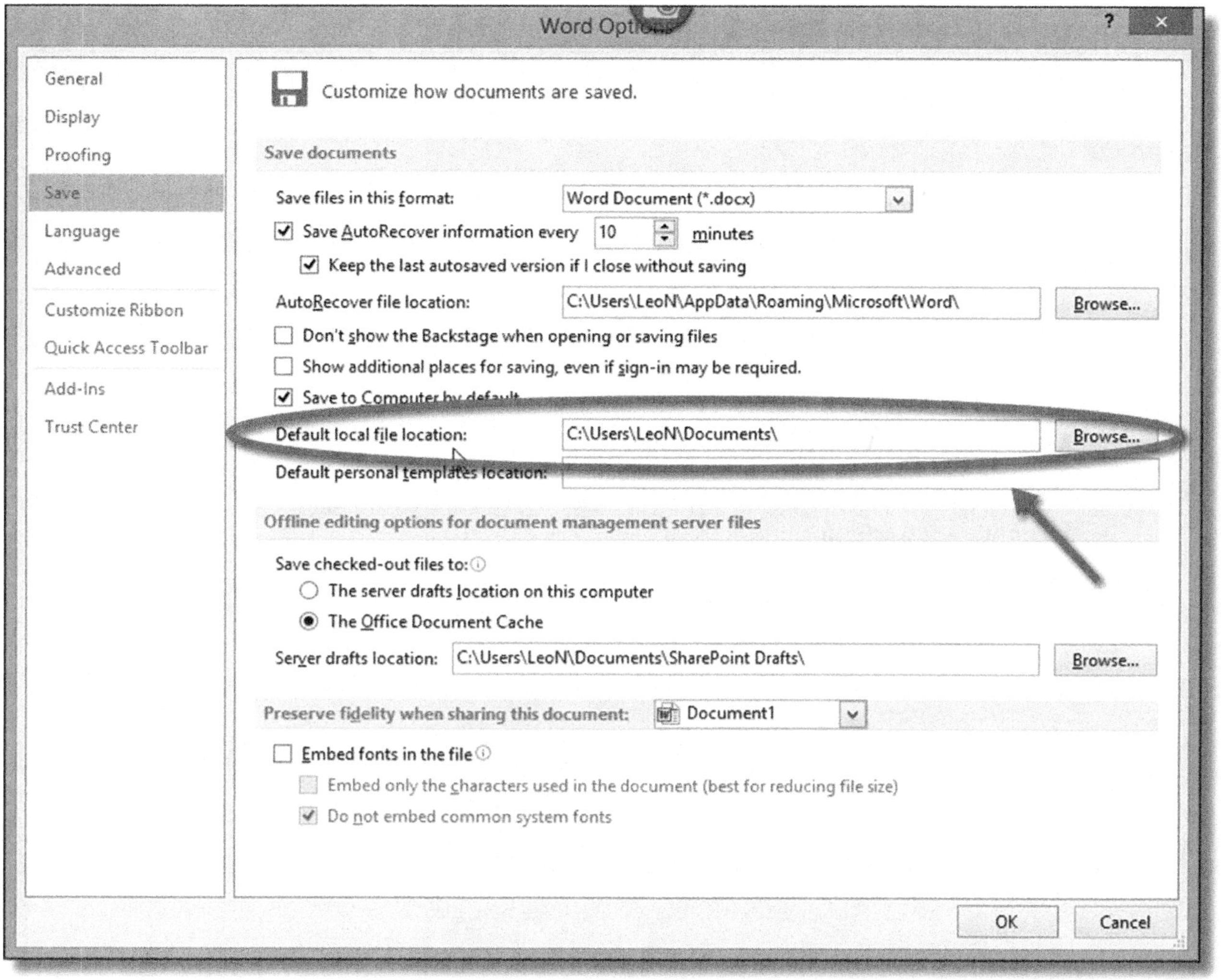

Change “Default local file location:” to be the full path to the Dropbox folder. In fact, it might be worthwhile to create a folder *within* your Dropbox folder for just this purpose. I created a “Documents” folder within my Dropbox folder, and then changed the Word default folder to C:\Users\LeoN\Dropbox\Documents:

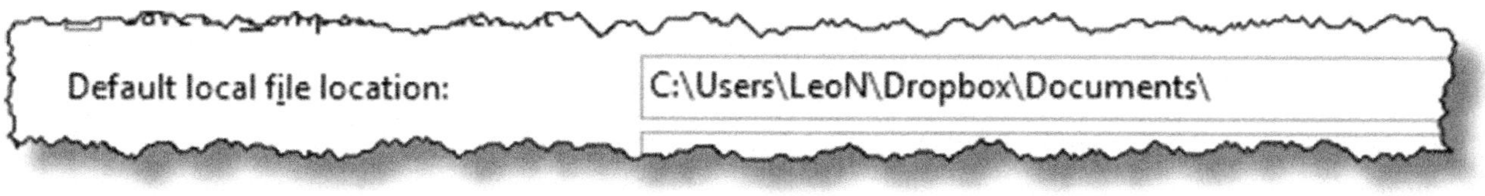

Now when you create a new document in the default location that document is automatically replicated to your Dropbox account online.

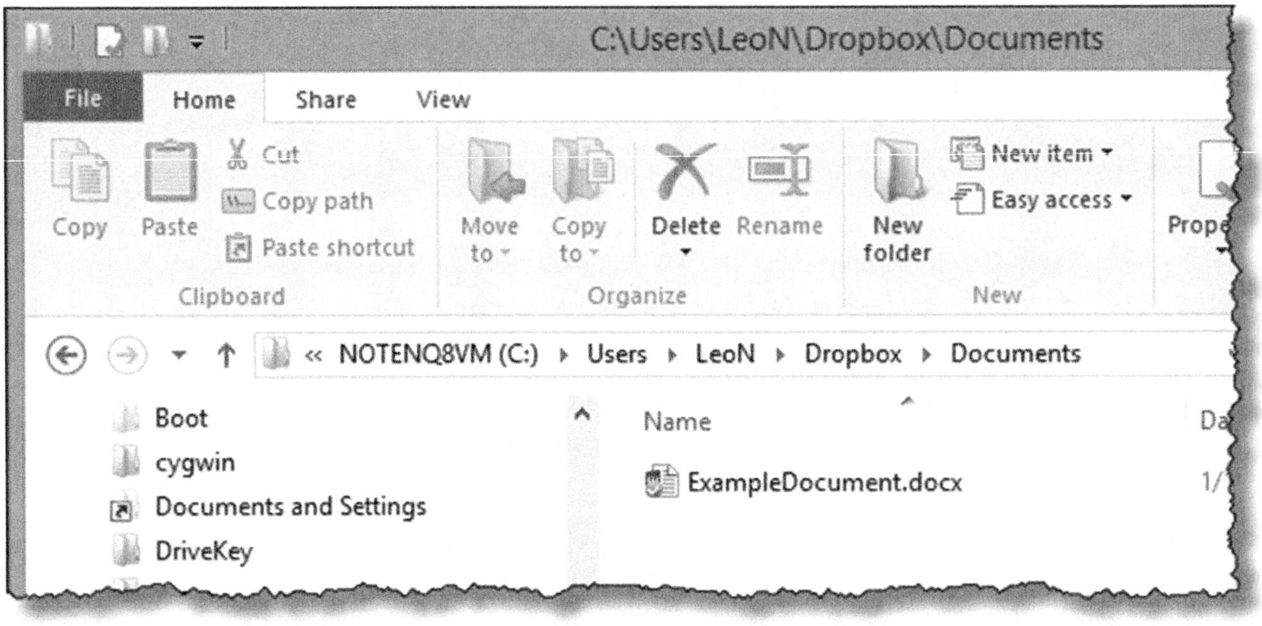

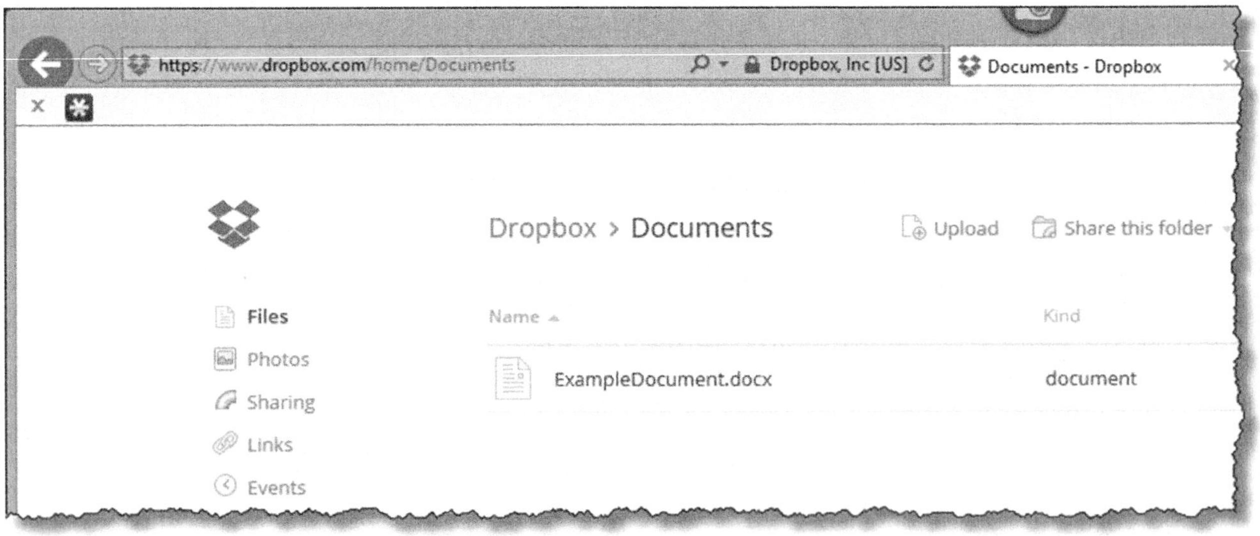

In fact, *every time you hit "Save"* in Word, the document is uploaded to Dropbox. You can always download the most recent copy from the Dropbox web interface.

Version history

As I've said, the big selling point of Dropbox, and utilities like it, is that you can install it on multiple machines, and the files you place in your Dropbox folder will be synchronized across all of them.

Dropbox online also includes version history. Quoting the Dropbox site: "By default, Dropbox saves all deleted and previous versions of your files for 30 days." Thus, if you're using Dropbox as I've outlined here, and you make a number of changes, or accidentally delete a file in the Dropbox folder on your hard drive, not to worry: you can still use the Dropbox online interface to retrieve previous copies of the file.

Using Dropbox for work in progress

My example illustrated using Dropbox with Microsoft Word, assuming that's where you do most of your work. In reality, any program that saves data to your computer can be set to default to your Dropbox folder.

I use Dropbox folders for my Word documents, text files, pictures... just about anything that's a "work in progress". (In reality, since I also use this technology on multiple machines, I keep much more in these folders, simply so they're replicated across all my machines.)

Just Do This:

Use Dropbox or programs like it to provide nearly continuous backups for the files you work on between backups. To make this easy, do all your work on files stored within the Dropbox folder on your machine. To make that happen automatically, change the default document location in the applications that you use to be your Dropbox folder, or a sub-folder within it.

5

Use BoxCryptor Classic to Protect Your Data in the Cloud

One of the very first comments I get whenever I talk about Dropbox, or even cloud storage in general, is the issue of privacy. Many are concerned that by placing your sensitive (or even not-so-sensitive) documents into an online storage provider you run the risk of that data being exposed to hackers, government agencies, and just generally people you don't want seeing your data.

At a minimum, almost all online storage providers have the ability to examine your data or hand it over to law enforcement, even if the provider has encrypted your data.

Hopefully, most of us will never have to deal with the law-enforcement scenario, but even the realization that a rogue employee at an online data storage provider could peek into what we keep online can cause concern. For some, it's enough of a concern to avoid using cloud storage at all.

The solution is simple: encrypt the data yourself.

Unfortunately, implementing that "simple" solution isn't always that simple or transparent, and can add a layer of complexity to online storage that some find just as intimidating.

BoxCryptor Classic is a nicely unobtrusive encryption solution and free for personal use.

The hidden issue

Online[9] storage is powerful when used properly. By "properly," I mean that the account used is set up with appropriate security, and that the data you might place online is still appropriately backed up somewhere else as well. Ignoring either of those items can lead to permanent data loss or worse.

Even with those basics covered, though, there remains a potential privacy issue. Unless you take additional steps of your own, the provider of that online service has the ability to view your data.

- Your data may not be encrypted, and is stored in view of anyone with administrative access to the service provider's servers.
- Your data may be encrypted by the service provider, but since it's encrypted by the service provider, it can also be decrypted by the service provider. As a result, those with administrative access still have the ability to view your data.

In reality, the folks at major online storage providers are professionals with no interest in snooping around in your data. Instances of the so-called "rogue employee" are rare, but still possible.

More legitimately, the service provider may be required to turn over your unencrypted data to law enforcement should the appropriate court orders be presented[10].

For whatever reason, you might consider all of this a problem. Fortunately, this problem has a solution: encryption.

The hidden cost of doing your own encryption

There's actually one good reason to allow your online storage provider to have the ability to decrypt your data: web access.

9 The over-hyped marketing term "cloud" is really nothing more than a replacement for online. What is often called "cloud storage" is nothing more than storage provided by online services.

10 Depending on the laws in your locality, of course.

If your online storage provider encrypts your data when it's stored on their servers, they must be able to decrypt it to provide you the ability to access the data via a web interface. Dropbox, for example, allows you to log into your account from any machine and access the files stored in your account *via the web*.

If you encrypt the data yourself using a tool like BoxCryptor, the online storage provider can only get access to your data in its encrypted form. You'll need BoxCryptor on your computer, and of course, your password, to decrypt it before you'll be able to use that data.

Traditional encryption solutions

The idea here is simply that if you encrypt your data before it is uploaded to any online storage provider then they have no ability to decrypt it. You, and only you[11], control the access to your actual data.

Traditionally, that works this way:

- You have a file or set of files that you want to store online.
- You use a program such as 7-zip, Axcrypt, TrueCrypt, or similar to create a new file or files that contains encrypted versions of the files.
- You place those files into online storage.

Now, when you want to use those files on any machine where you don't have the original or want to make sure you have the most current copy, you need to:

- Retrieve the encrypted files from online storage.
- Decrypt the files.
- Make your changes.

11 And of course, anyone you might choose to share the password with.

Then, finally, to update the online copies of the files (if you made any changes), you would:

- Re-encrypt the files as you did originally.
- Place the encrypted file or files into online storage.

As you can see, that's a lot of work just to update, for example, a single file.

That's where BoxCryptor comes in.

The BoxCryptor encryption model

When using BoxCryptor Classic you designate a folder on your machine as a special container for your encrypted data.

You never access the contents of this folder directly. Instead you "mount" that folder using BoxCryptor. In order to mount it, you'll need to specify your pass phrase.

Once mounted, another drive - I'll call it L: - appears on your machine. Anything written to drive L: is encrypted and written to the folder that you specified when mounting. Anything read from that drive L: causes the corresponding encrypted file in the source folder to be read and decrypted on the fly.

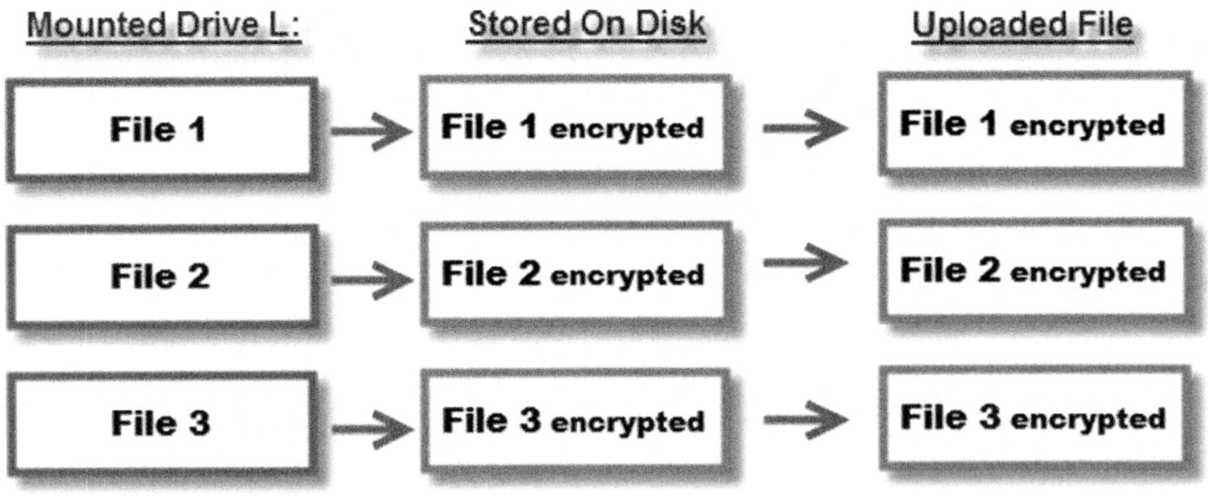

The files in the original source folder are always encrypted, as are the files uploaded to Dropbox. It's only when the folder is mounted in BoxCryptor that the files are visible in their decrypted form in the virtual drive.

Now your data is encrypted when it is uploaded to cloud storage, but you can easily see and work on it in an unencrypted form from drive L:.

Download and install BoxCryptor Classic

Download Boxcryptor Classic from the Boxcryptor web site[12].

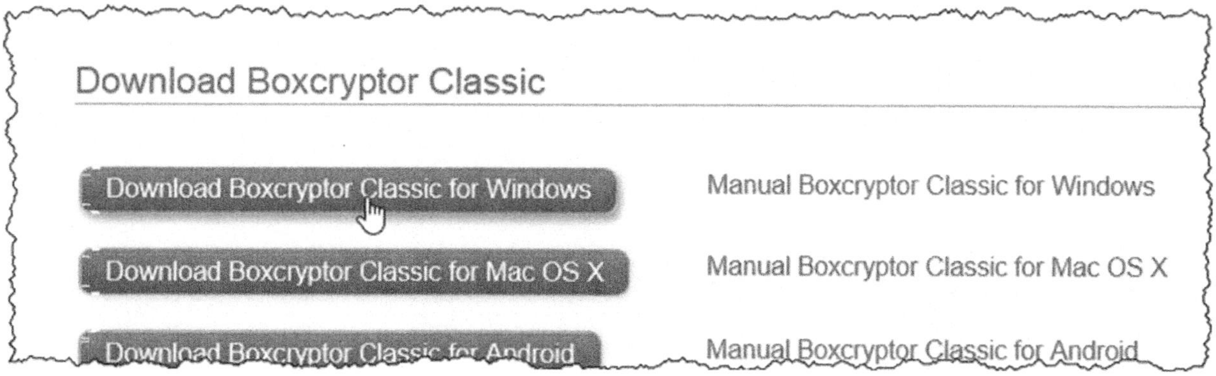

12 http://go.askleo.com/boxcryptor

The setup process is straightforward: simply follow the prompts to install the product.

Once you click **Finish** on the last set-up page, you'll be given the option to create a new Boxcryptor classic folder, or open an existing one.

Since we're just starting with Boxcryptor Classic, we'll create a new folder. If you later install Boxcryptor Classic on additional machines to access the same data, then you'll choose to open an existing one.

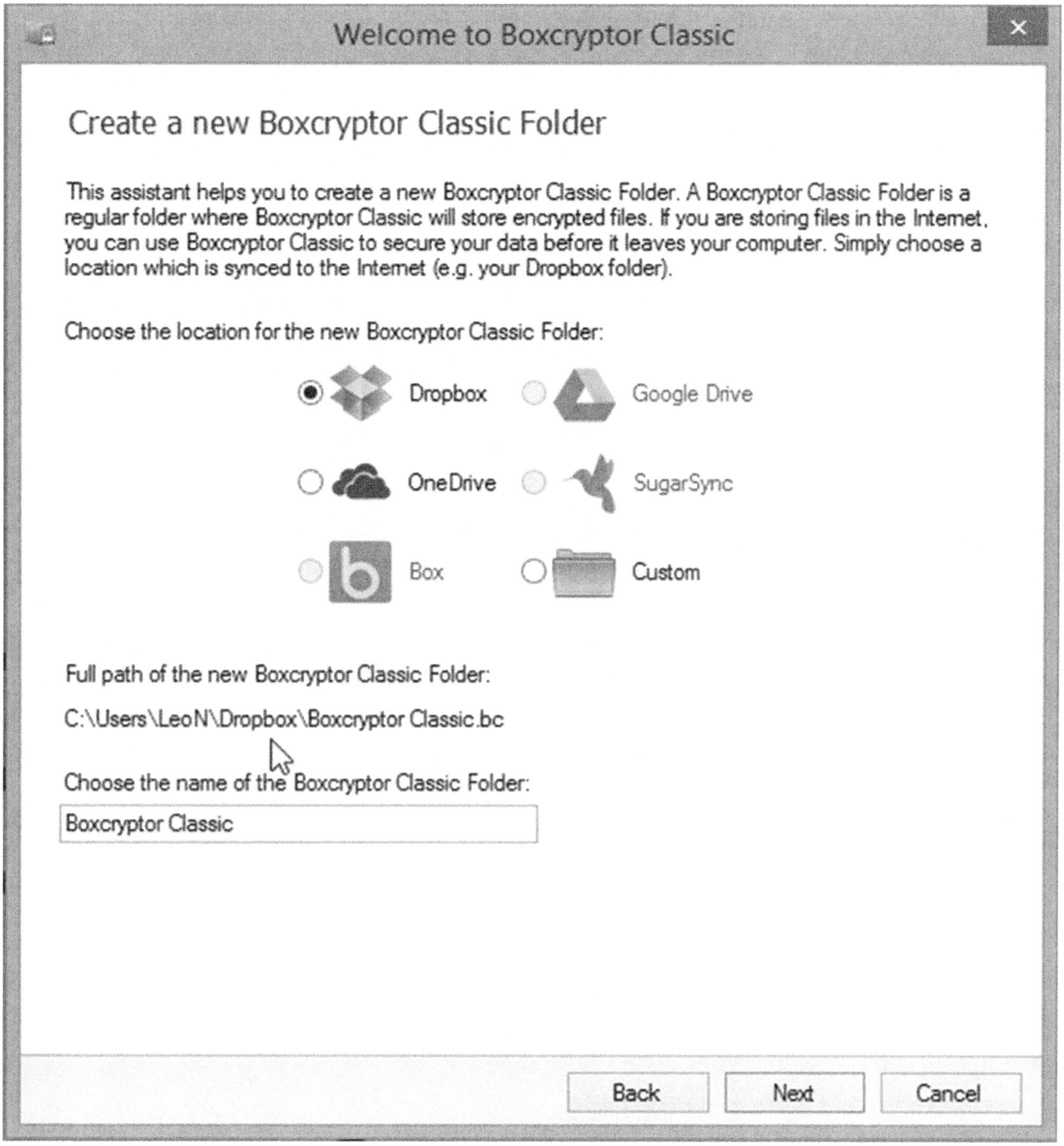

Boxcryptor is aware of several common cloud services, including Dropbox. All this does is allow Boxcryptor to locate the folder on your machine currently managed by that service. Since we installed Dropbox in the previous chapter, we'll select that here. Box-Cryptor correctly identifies the Dropbox folder on my example machine as:

C:\Users\LeoN\Dropbox

It also offers to create a folder therein to hold the encrypted data:

C:\Users\LeoN\Dropbox\Boxcryptor Classic.bc

Click **Next** to accept the defaults on your machine.

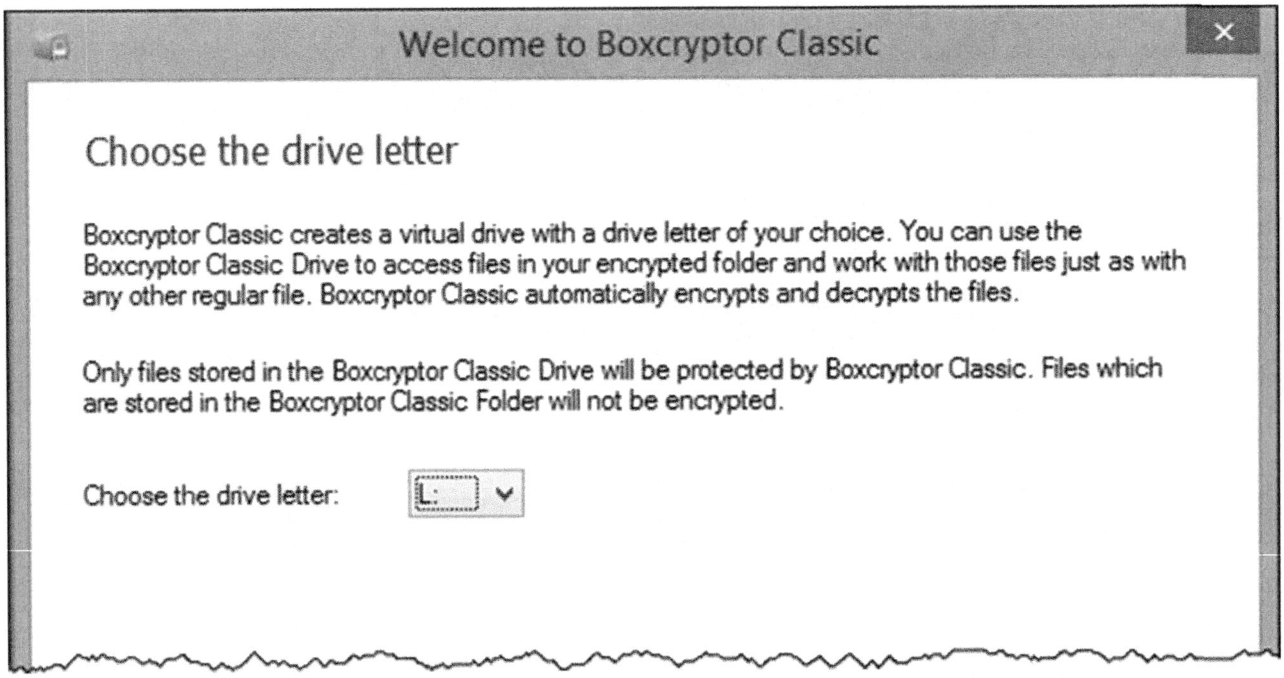

Boxcryptor then asks what drive letter you want the unencrypted data to appear under. I've selected "L:" in the example above, but you can choose whatever drive letter is available on your computer.

Click **Next**.

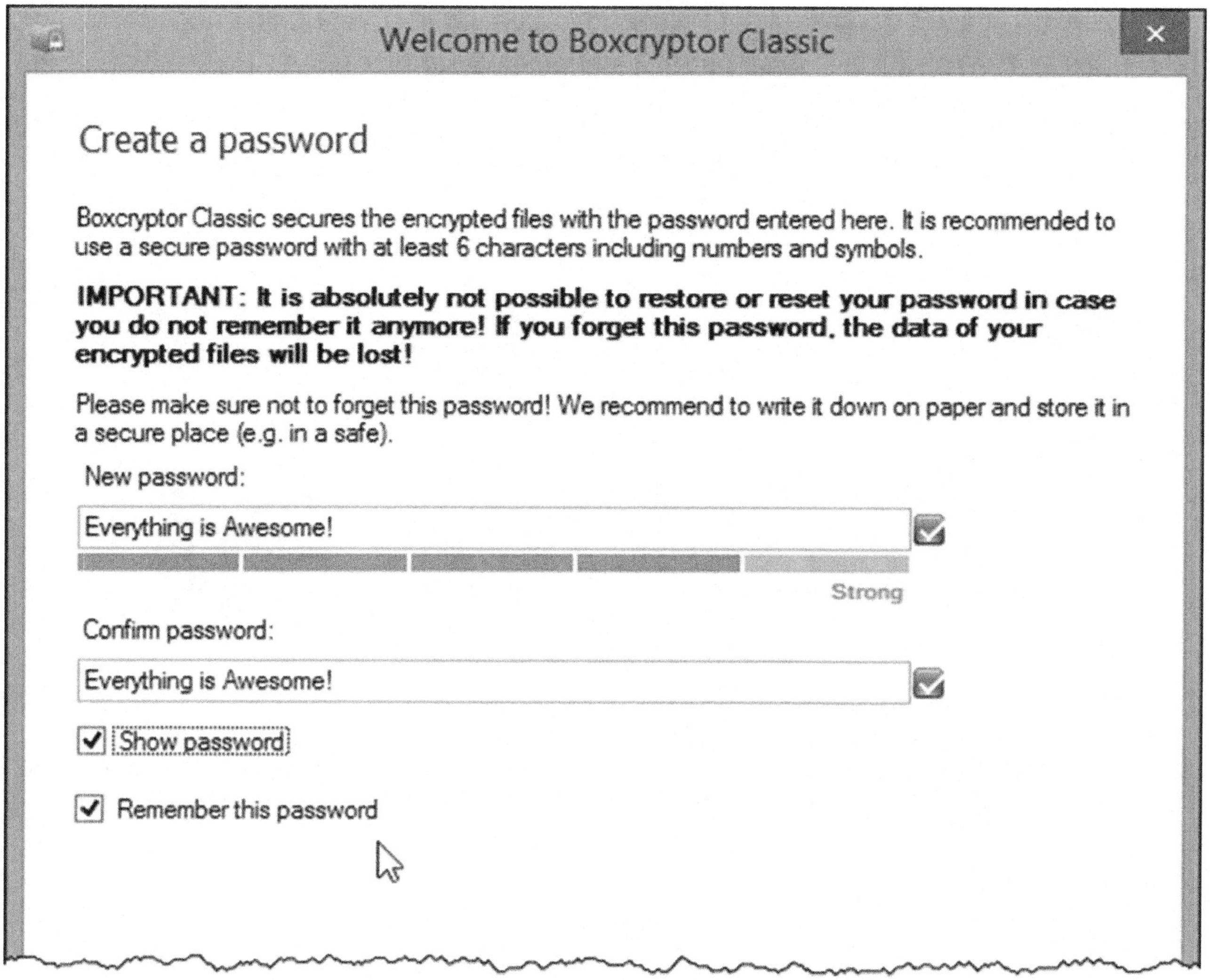

Now it's time to specify the passphrase that will be used to encrypt your data.

Note that I said pass*phrase*. This is a case where you want something exceptionally secure and much longer than your traditional password. The good news is that length trumps complexity, so we can actually use an actual, memorable, phrase.

In the example above, I've selected "Show password" so that you can see what I've entered. That's a good approach to making sure you're entering what you think you are. If, for any reason, you're wrong, and you cannot replicate this password, you won't be able to access your encrypted data.

The passphrase I've used is actually pretty reasonable. You should choose your own, of course, but "Everything is Awesome!" includes spaces, upper and lower case, a special character and is 22 characters long. Plus, it's easy to remember.

I'll talk more about "Remember this password" in a moment.

After choosing your own passphrase, click **Finish**.

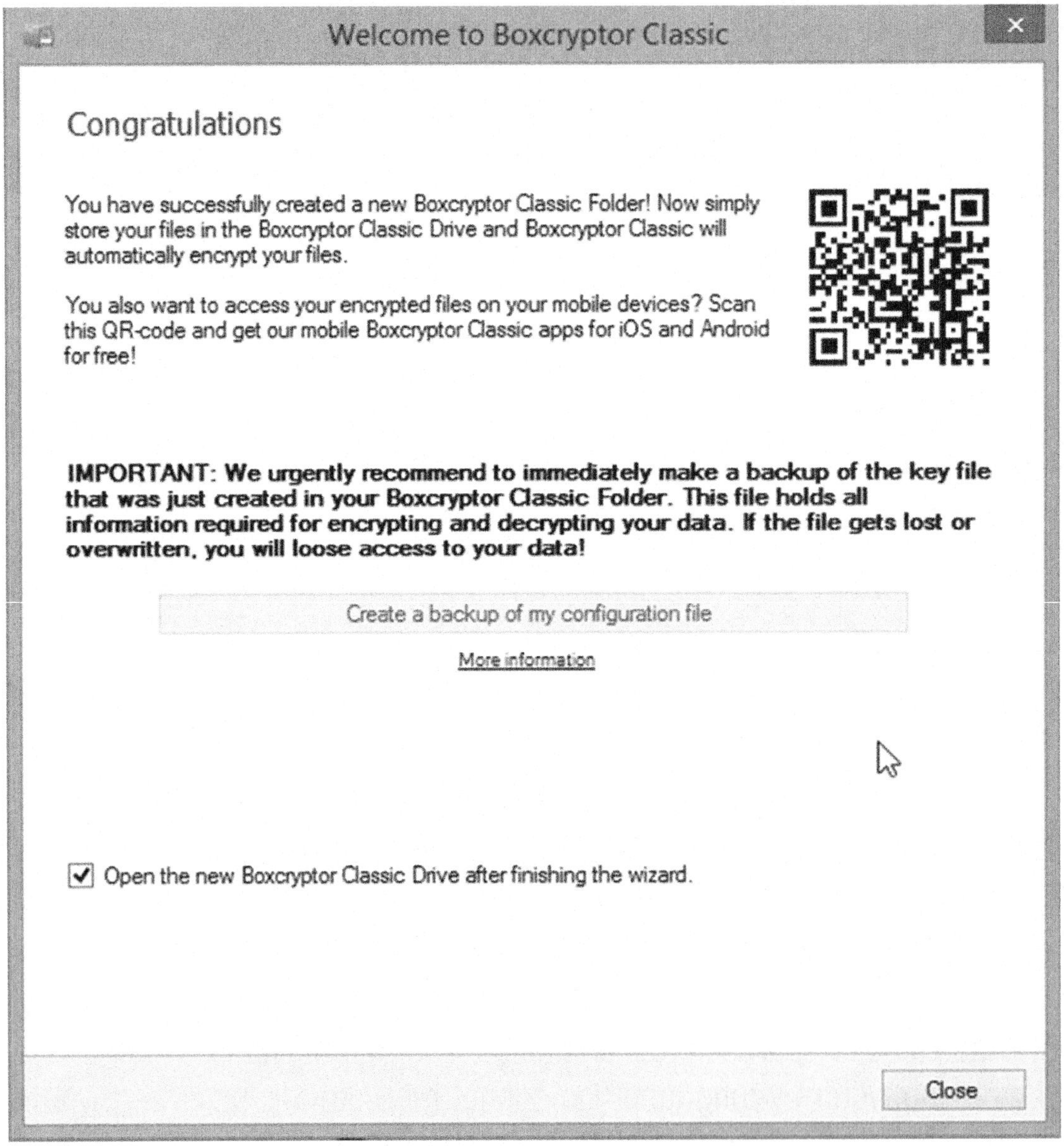

In addition to displaying a QR code you could use to connect Boxcryptor on a mobile device, this final page urges you to back up files critical to your Boxcryptor installation. Even though we've set up our monthly image backup, I still strongly recommend you do this, and save the resulting backup files in a safe place. They're critical to Boxcryptor being able to decrypt your files.

Click **Close** and you're done. Boxcryptor Classic is installed and configured on your machine, protecting a folder within your Dropbox folder.

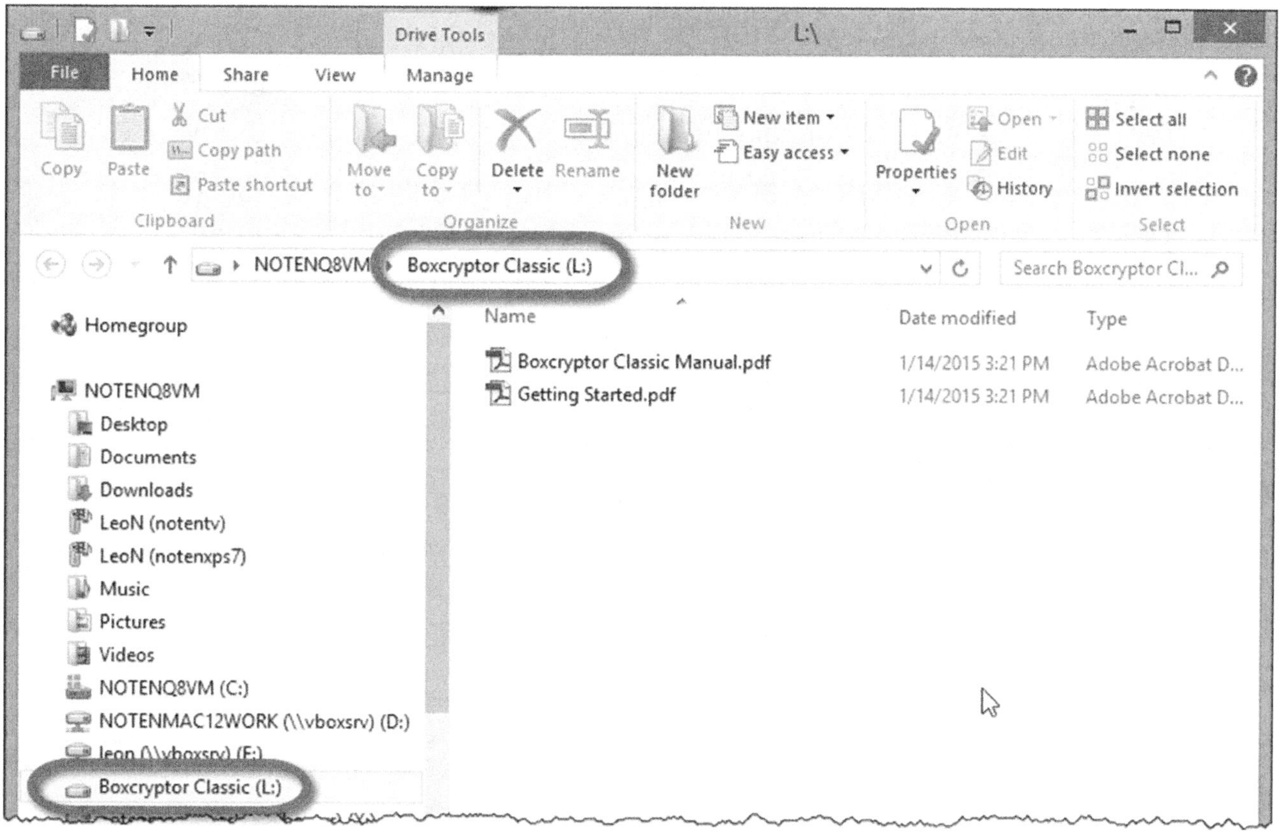

In fact, you'll find your first encrypted files therein already: the Boxcryptor manual. When accessed from drive L:, as shown above, the files are unencrypted and valid PDF documents. However, if you try to access them directly from C:\Users\<username>\Dropbox\Boxcryptor Classic.bc you'll find that the files are unreadable: they're encrypted.

Mounting Boxcryptor folders and remembering pass phrases

You'll note that in the setup process above I elected to let Boxcryptor remember the passphrase. That means that whenever Boxcryptor is run it will automatically mount drive L: using that passphrase, and your unencrypted data will be readily available.

Isn't that a security risk?

Yes and no. Here's my recommendation:

- On your home machine, which is relatively secure, let Boxcryptor remember the pass phrase and automatically mount the Boxcryptor folder and drive.
- On a portable machine, or any machine that you consider *not* secure, don't. Instead, when Boxcryptor loads at start up it'll do nothing. You'll need to right click on the Boxcryptor icon in the notification area of your desktop, select **Mount**, and enter the passphrase associated with the folder. Only then will the unencrypted data, and the drive ("L:" in the examples above) appear. When you are done, you can right click it again and select **Unmount** to re-secure your data.

Regardless of which you choose, only the encrypted data is uploaded to Dropbox. Your data on the Dropbox servers is secure, and that's our primary purpose.

It's for more than PCs

Much like many online services similar to Dropbox, BoxCryptor supports multiple platforms.

BoxCryptor is available for:

- Windows
- Apple OSX
- Android
- iPad and iPhone

What that means is that you can continue to share your documents across all the platforms and devices supported by your online storage provider, but now you can easily encrypt the data you share.

BoxCryptor or BoxCryptor Classic?

BoxCryptor changed their flagship product a while back. Among other things it requires a BoxCryptor account, and apparently some of your information is stored in their online storage.

That's somewhat counter to my desire to keep things secure and completely private.

Thus, my comments and recommendation relates specifically to the original version of their product, BoxCryptor Classic. While I'm sure it's a fine product, I do not recommend the current incarnation of the flagship BoxCryptor product, referred to simply as *BoxCryptor*.

It's free for personal use, but...

I actually recommend you spring for the personal license.

Besides better support, it includes a feature that I suspect many people might want: filename encryption.

As you saw above, an example document:

L:\MyPrivateInformation.docx

would be saved as:

C:\Users\LeoN\Dropbox\Boxcryptor Classic.bc\MyPrivateInformation.docx

In other words, the name of the file remains visible.

For many, that might not be a problem, but for others, names of files (and folders) can represent an unexpected way that sensitive information can leak out, even if the contents of those documents are encrypted.

When you purchase the "Unlimited Personal" license, file names are instead stored encrypted. For example, my example document might appear as:

C:\Users\LeoN\Dropbox\Boxcryptor Classic.bc\gVbJ27u6-VMQ

Only when successfully mounted will the file names once again appear unencrypted (on drive L: to continue the examples above).

Of course, there's a commercial license as well.

Just Do This:

Download and install Boxcryptor Classic, and configure it to operate on a folder within your DropBox folder, as I did in the example above. Now you can feel safe that any documents you place in a mounted Boxcryptor folder will be safe from prying eyes in the cloud.

6

Automatically Backup Smartphone Photos

I'm going to take a small departure from the PC-centric discussion so far, and talk for a moment about using your smartphone.

Specifically, since we just installed Dropbox, I want to show you how to install and use Dropbox as an automatic way of backing up the photographs you take using your smartphone.

I'll use my Android-based Samsung Galaxy Note for these examples, but the concept applies to just about any smartphone running iOS or Android.

Install the Dropbox app

Go to the Google Play store on your mobile device, and search for Dropbox.

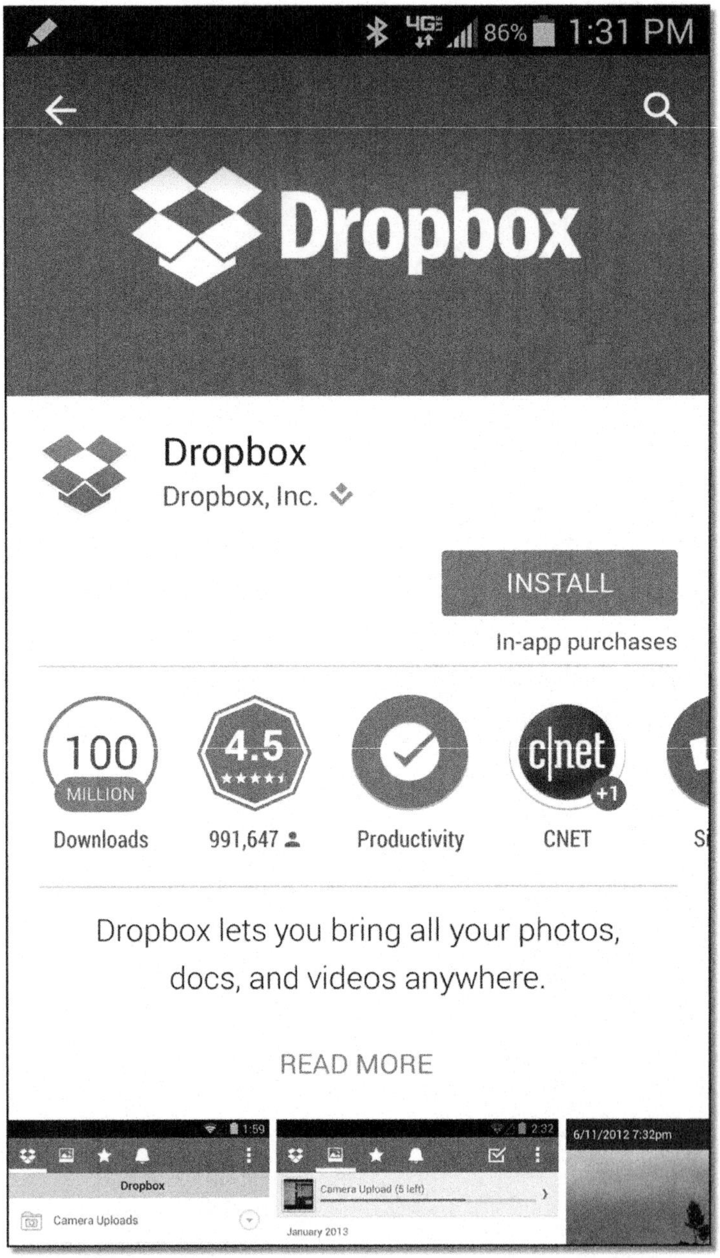

Tap on **Install**. The installer will confirm the permissions used by Dropbox.

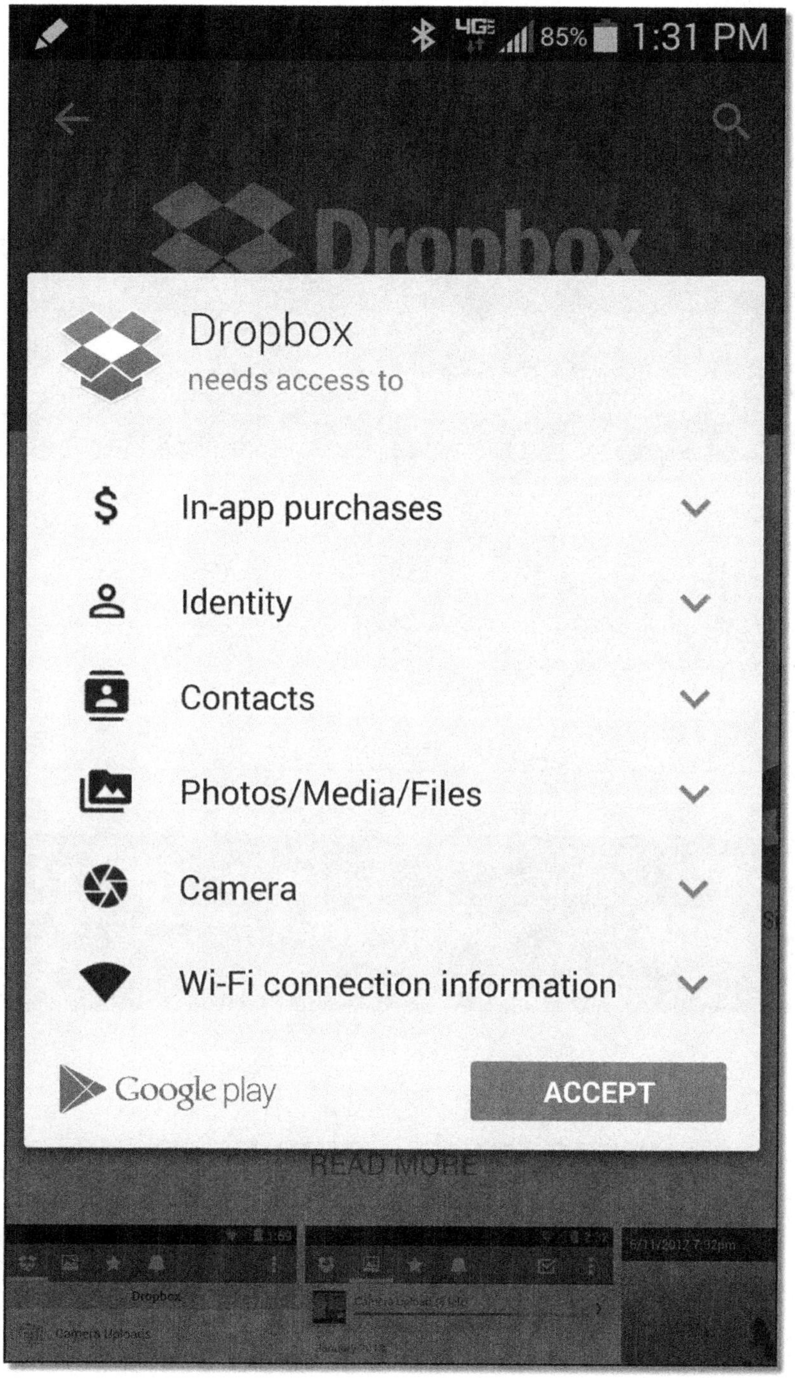

Tap on **Accept**.

That's it! Dropbox is installed on your device. The installer will return to the Google Play Store page, and the "Install" button will change to "Open". Tap **Open** to start Dropbox.

Configuring Dropbox

In a previous chapter you created a Dropbox account and installed Dropbox on your PC. Now, in the mobile app, enter your account credentials.

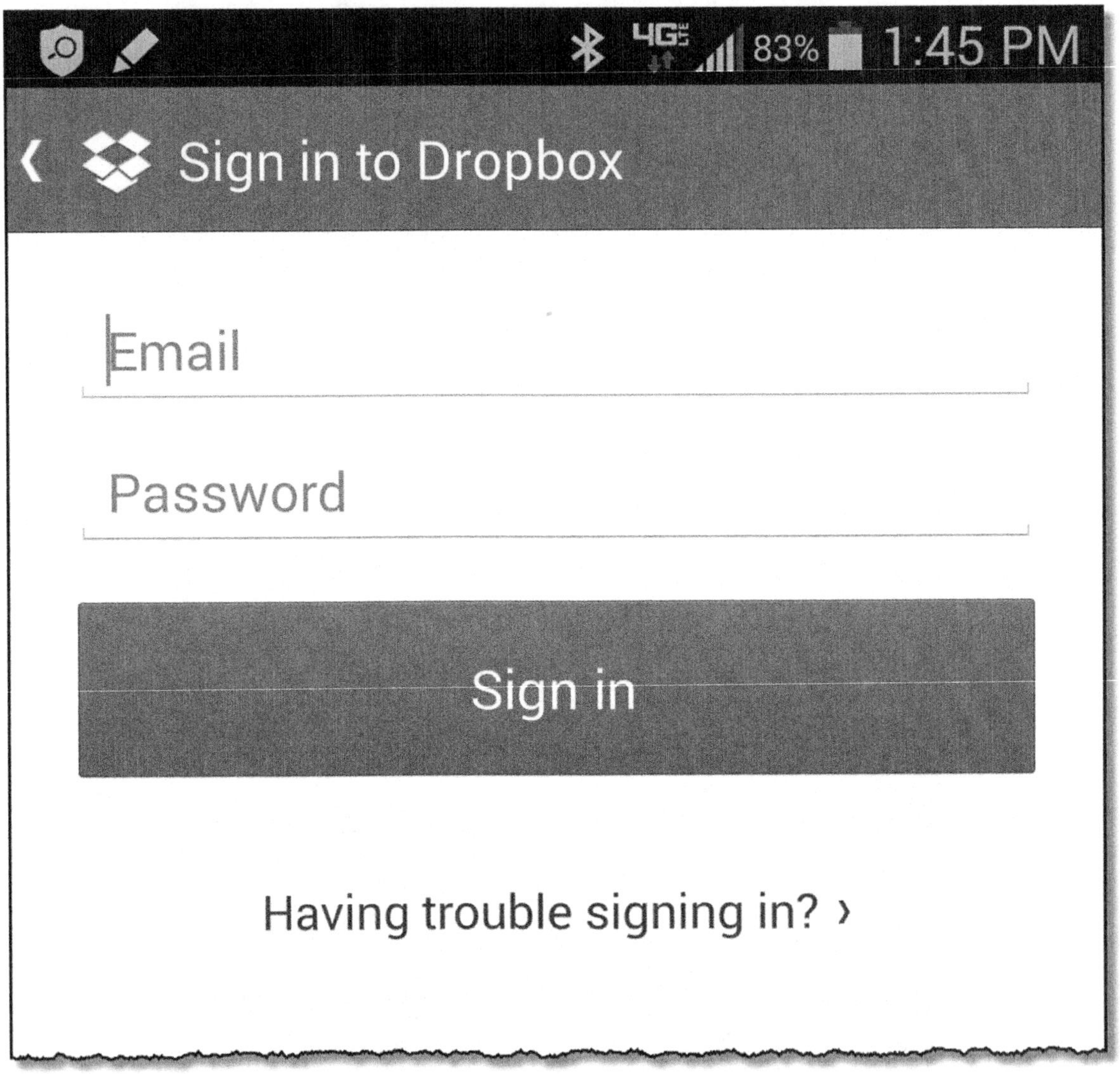

Tap on **Sign in**.

One of the reasons I like Dropbox over some of its competitors is that it supports two-factor authentication[13]. Since I have that enabled, Dropbox next sends me a text message with my security code. After I click **Sign in**, Dropbox presents a screen into which I must type the code that was just sent.

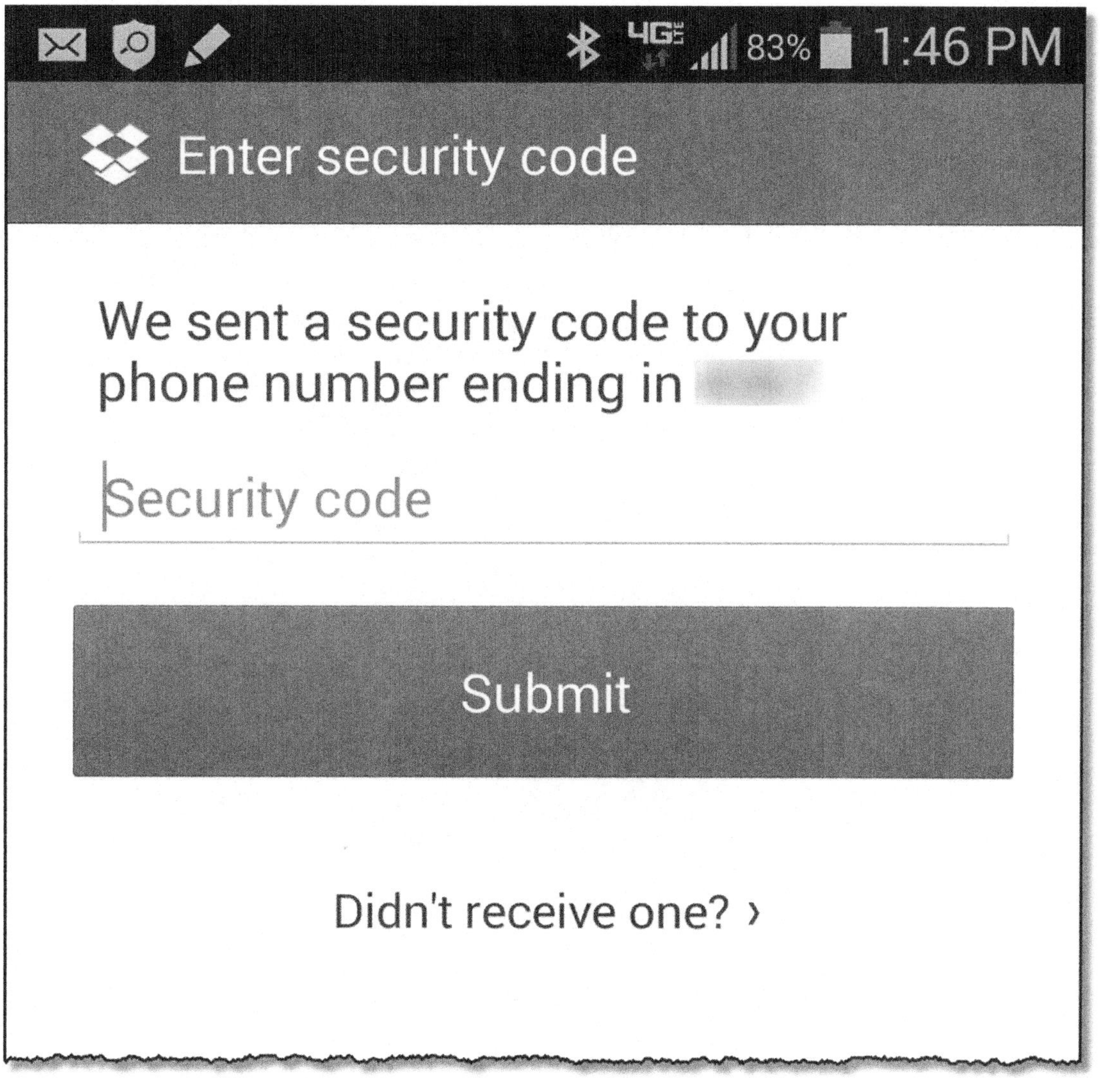

13 You can learn more about two-factor authentication in this Ask Leo! Article: https://askleo.com/16401

It may seem a little odd to get the text message on the same device to which I'm installing Dropbox. This prevents someone else from signing in to my account on their own Dropbox app, because even if they somehow know my password, they won't have access to the text message on my phone. Once you've signed in successfully on your mobile device, you won't need the second factor code again unless you explicitly sign out.

Tap on **Submit**, and Dropbox is good to go.

Enable auto-upload

Before letting you move on, Dropbox offers the automatic upload feature.

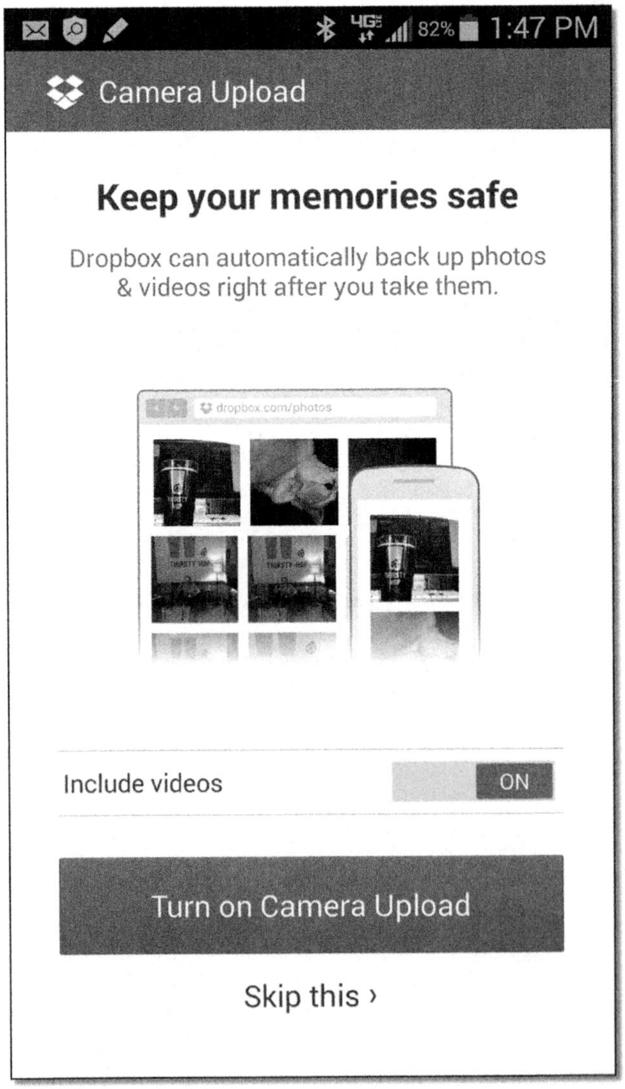

This is why we're here. Turn on "Include videos" if you like (as I did), and then tap on **Turn on Camera Upload**.

Dropbox will begin uploading and backing up the photos you've already taken. From now on, however, photos you take will be uploaded almost immediately.

Uploaded where?

Accessing your uploaded photos

Dropbox creates a new folder called: "Camera Uploads".

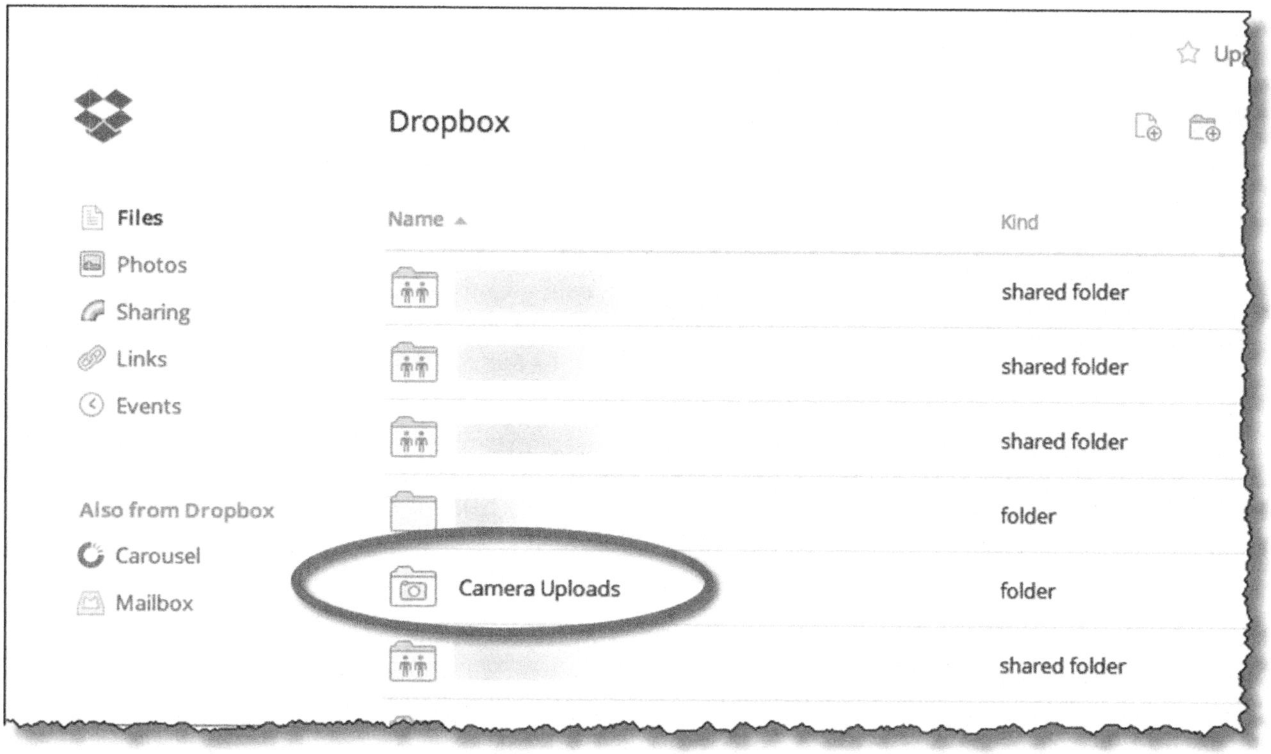

Examine the contents of that folder, and you'll find all your photos[14].

14 The observant reader might notice that the thumbnails in that image reflect the other images on this page. The screen shot capability of the Samsung mimics a camera, so as each image was captured, it was automatically uploaded to my Dropbox account, and then downloaded to my desktop machine, where I was writing this article.

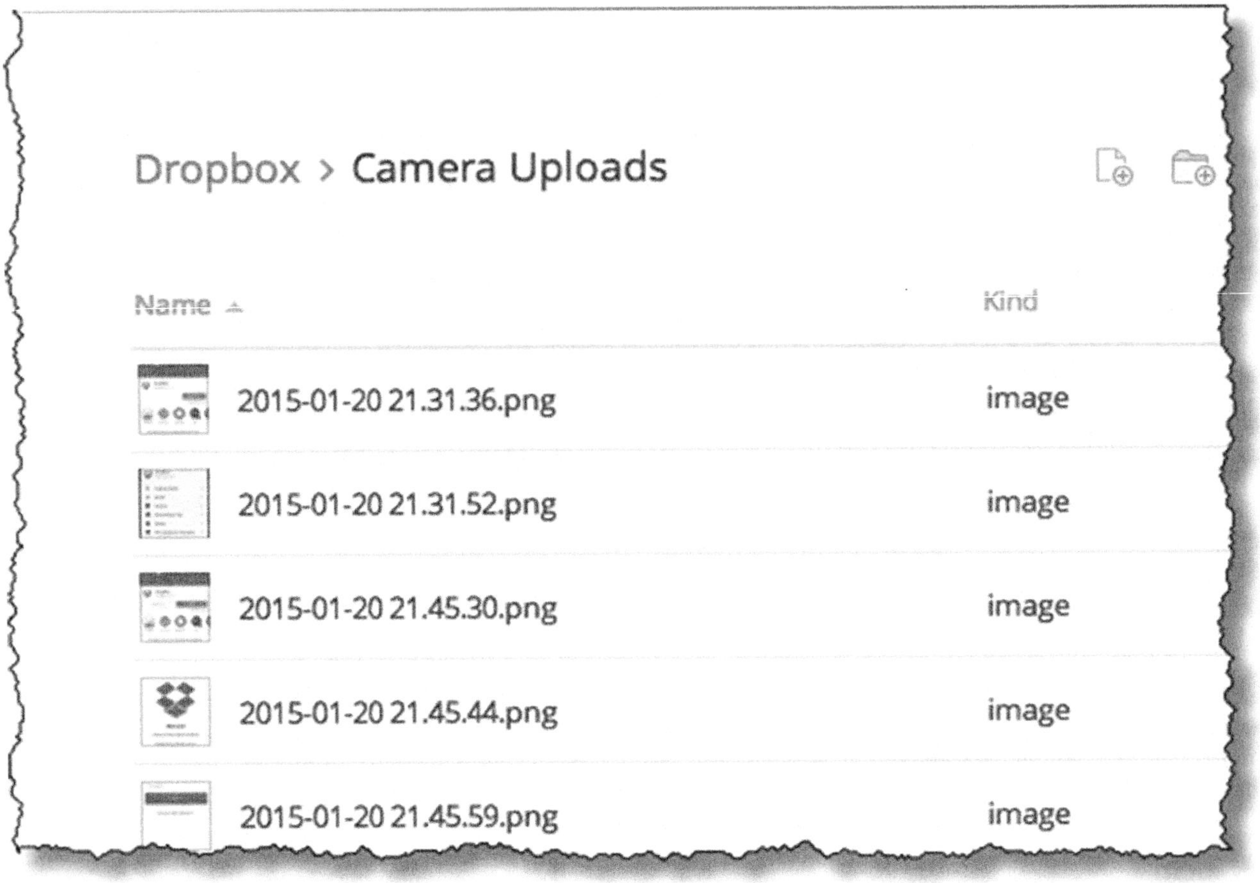

If you have Dropbox installed on more than one device, and you're signed in to the same account on each device, the photos will download automatically to all those devices as well.

Controlling the upload

The Dropbox app's Settings menu includes several options which allow you to control exactly what and when Dropbox uploads.

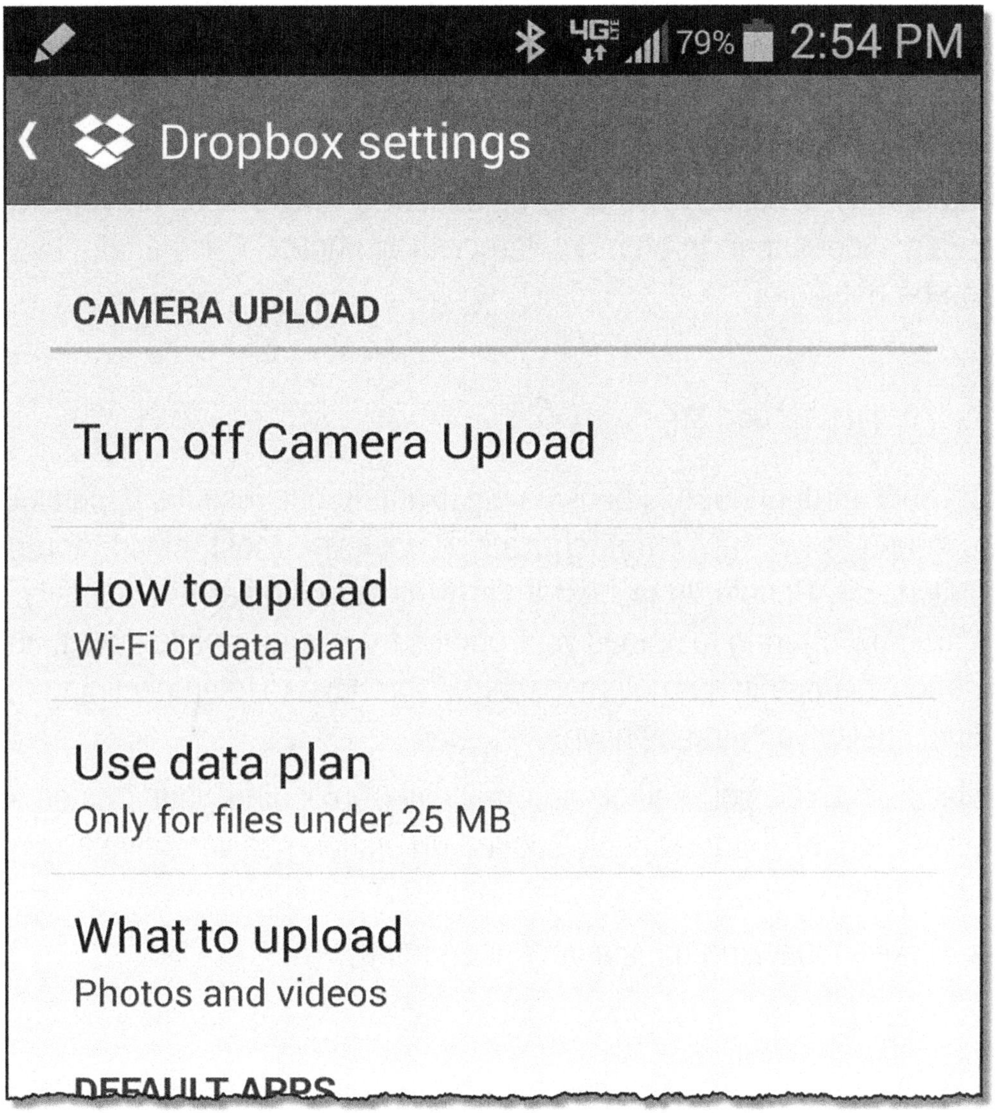

Naturally, you can turn the entire upload feature off. Unless you have another app that's doing the same thing, or have some other strategy for backing up your photos, I don't recommend this.

Because data transfers on your mobile provider network can count against your data plan or limit, and different people have different plans with different limits, you can choose to have Dropbox upload only when connected to a Wi-Fi network. I'm allowing it to upload no matter how it's connected.

If using your mobile provider data plan is enabled, you can still control how much is uploaded by setting a specific size limit on uploaded files. I've elected to allow anything up to 25 megabytes in size to be uploaded when connected to my mobile provider's network; anything larger than that will upload only when I'm connected via Wi-Fi.

Yet another way to manage what is uploaded is to allow, or prevent, videos from being uploaded. Videos tend to be much larger than photos. Once again, I've elected to allow both to be uploaded.

Dropbox or another service?

I've used Dropbox throughout this book as my example because it's popular, ubiquitous, robust, and supports two factor authentication. However, most, cloud storage applications from Microsoft, Google, Amazon, etc. provide very similar functionality. In fact, so many apps are now offering to upload your photos from your mobile device automatically that it's almost becoming a problem, because you have to keep declining the offers to replicate functionality you already have.

Regardless of which you select, automatically uploading your photos and videos means that you're almost immediately backed up, as long as you have an internet connection of some sort.

And as we know, backing up is a very good thing.

Just Do This:

If you have a smartphone, install Dropbox and enable its ability to upload the photos you take automatically. Not only will that make accessing them from your PC easier, it provides immediate backup.

Speaking of backup, there's something else I want to make sure you back up.

7

Backup Your Online Email with a Desktop Email Program

Many years ago, shortly after Ask Leo! began, I received a panic-stricken email from an individual whose account had been hacked. He had lost all access to the account and everything in it. His panic stemmed from the fact that, for whatever reason, the only copy of his master's thesis had been in that account.

It was gone, and there was no hope of recovery.

Hopefully, you're not keeping the only copy of something as important as a master's thesis in your online email account. That's wrong on several levels. But I'm guessing there are things in your account that you never want to lose, such as photos, correspondence, or other things you've exchanged in email.

The problem is, of course, that if it's only in your email account, it's not backed up.

Let's fix that. Let's back up your email.

Our example: Outlook.com and Hotmail

For this chapter, I'm going to use a Hotmail account - askleoexample@hotmail.com. I'll continue to access that account as I usually do, using the Outlook.com website, but we'll back it up to my PC using Thunderbird.

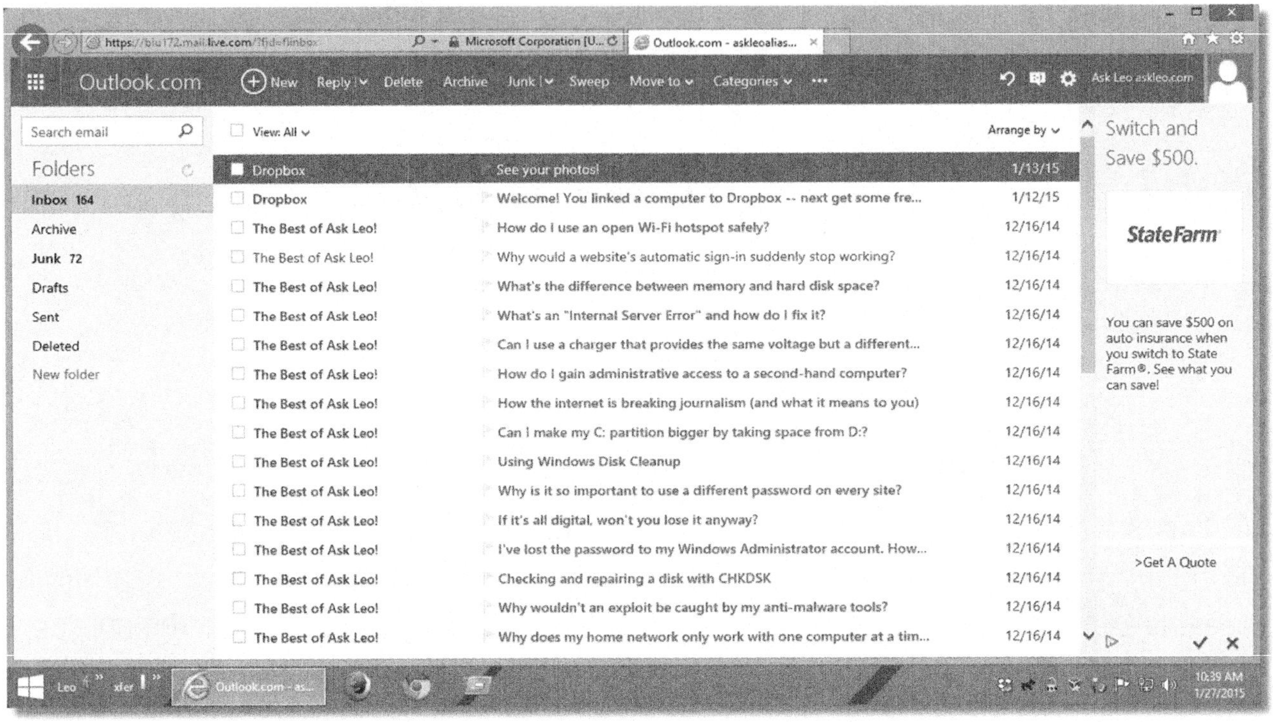

The technique I'll use applies to almost all online accounts, including Gmail, Yahoo! and others. For this technique to work, the account must support a protocol called "IMAP[15]". The good news is that IMAP is a common approach used to access email, and almost all email services support it.

In this example, I'll use Thunderbird as my desktop email program, but any desktop program that supports IMAP can back up your email, including Windows Live Mail, Microsoft Office's Outlook, and more.

15 Read more about IMAP at https://askleo.com/5965

Install Thunderbird

The first step is to download and install Thunderbird. Download Thunderbird from the official download site.[16]

At this writing, the official URL is based on https://www.mozilla.org.

Download and run the Setup Wizard.

16 It's critical that you download only from the official site. Do not get Thunderbird - or any software - from download sites or aggregators unless explicitly instructed to by the official site. Almost all download sites now add unwanted software in addition to whatever it is you're looking for.

The first choice is whether to accept the “Standard” settings or “Custom”.

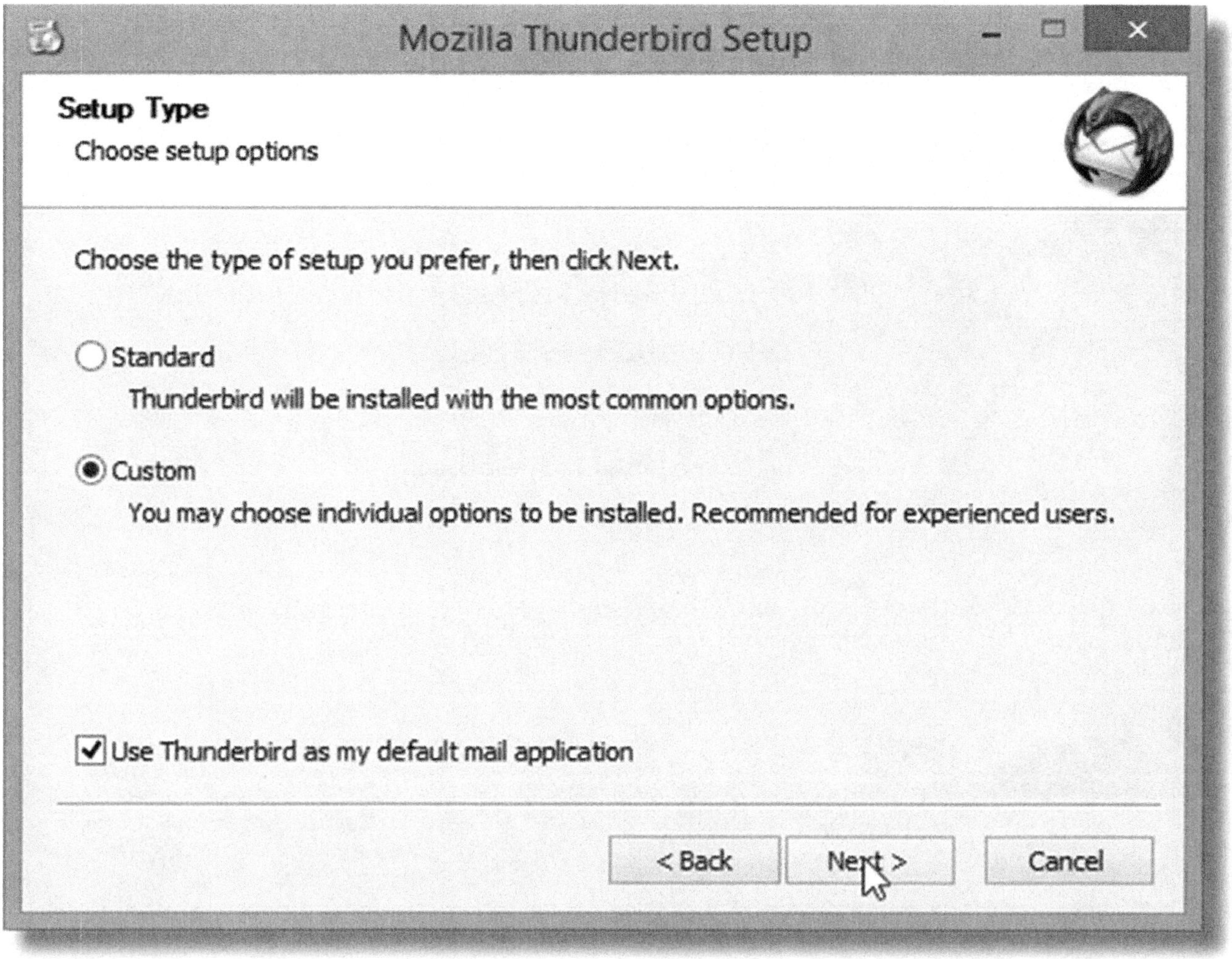

Always choose **Custom**. Even though we're going to accept all the suggestions made by the installer, choosing Custom is an important habit to get into to avoid malware and PUPs. Click **Next** through the subsequent settings, and accept:

- Create icons on the Desktop, Start Menu and Quick Launch bar.
- The installation location for Thunderbird.

If there are additional options, read them carefully to make sure you understand what it is you are agreeing to.

Click **Install** on the last page of the Wizard to install Thunderbird.

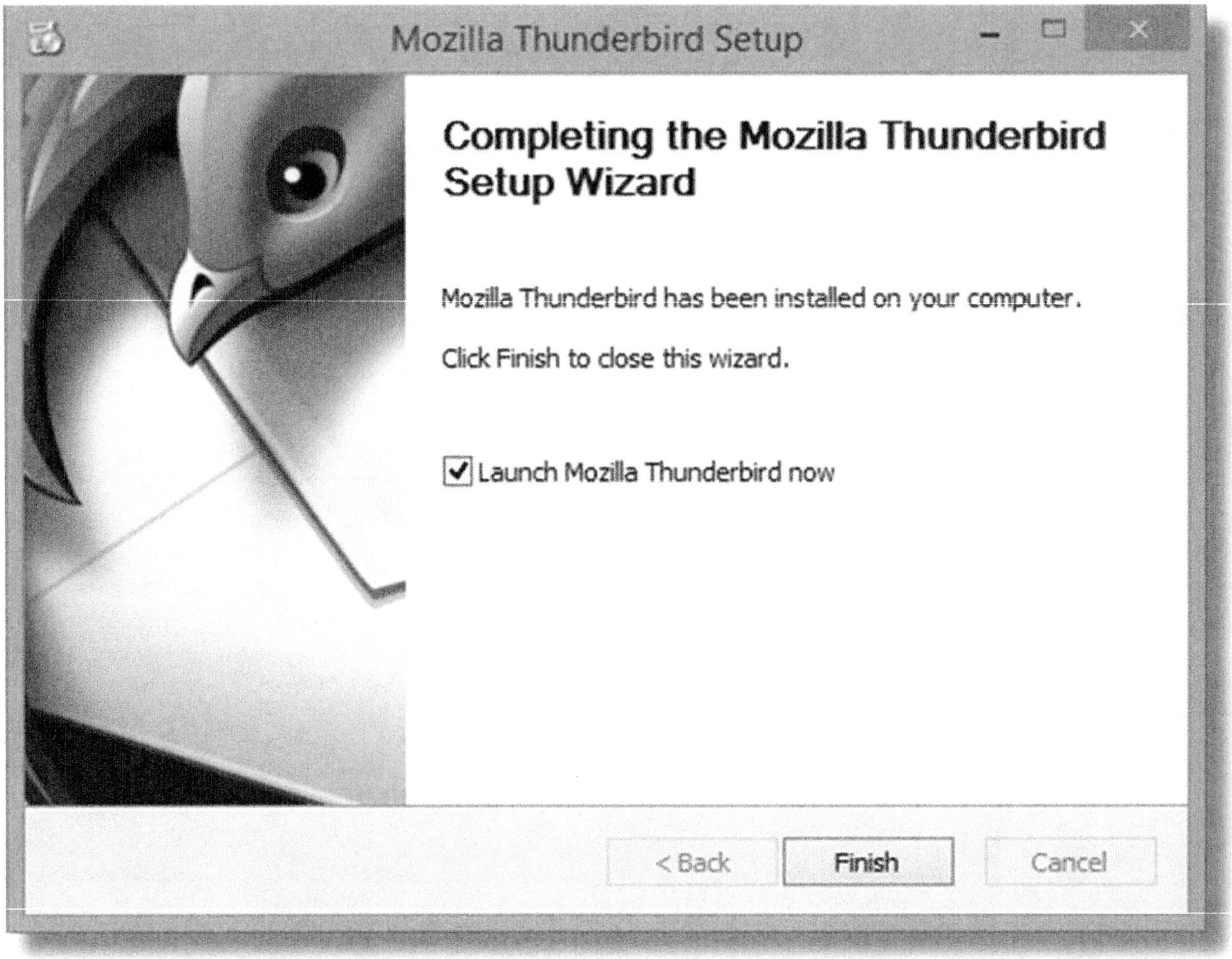

When the installation is complete, leave "Launch Mozilla Thunderbird now" checked, and click **Finish**.

When run, Thunderbird may offer to import settings from another mail program if one is installed.

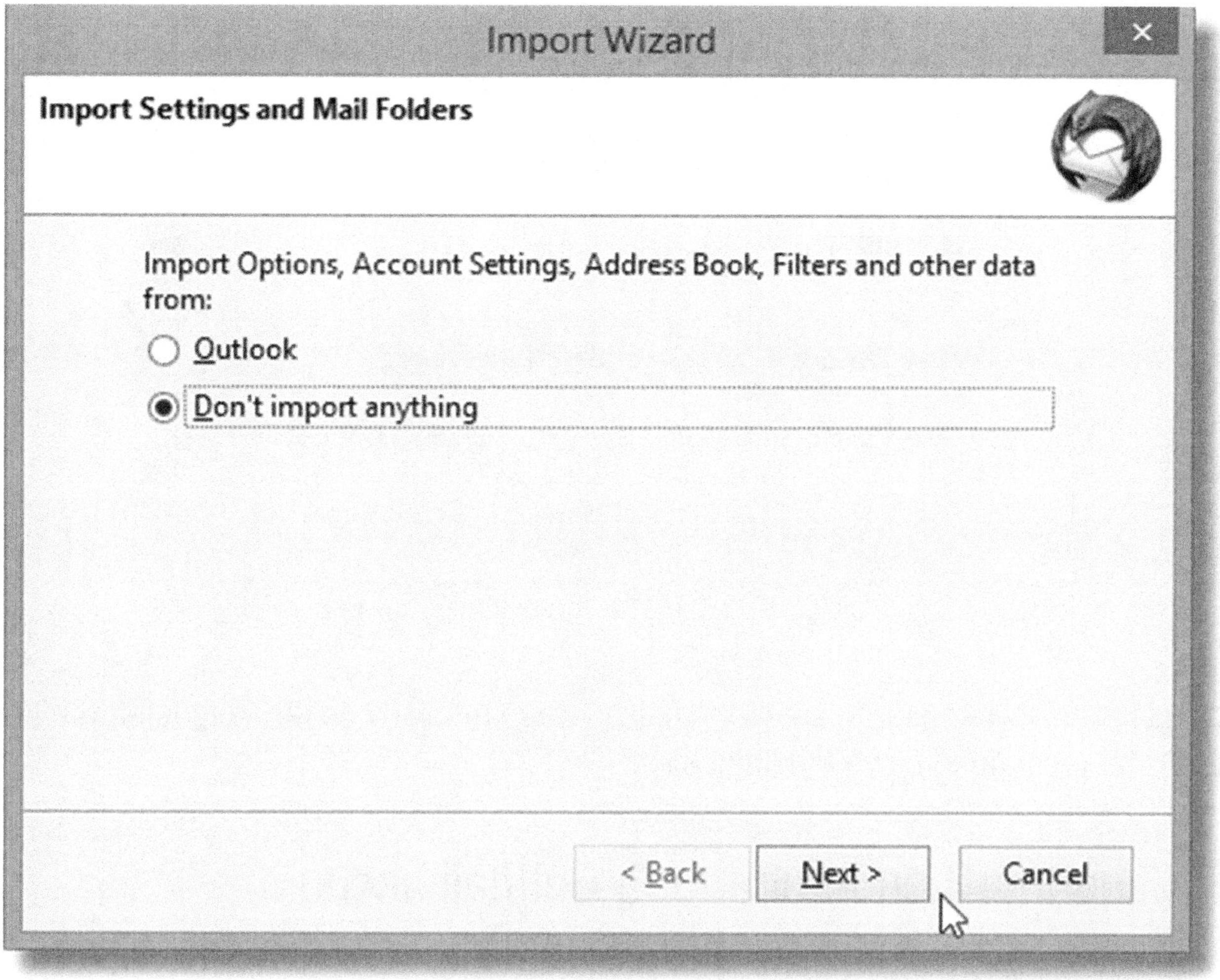

Since our goal here is simply to use Thunderbird for backup and we'll be configuring the account ourselves next, there's no need to import anything.

Thunderbird will then ask if you want it to be the default program for mail and feeds.

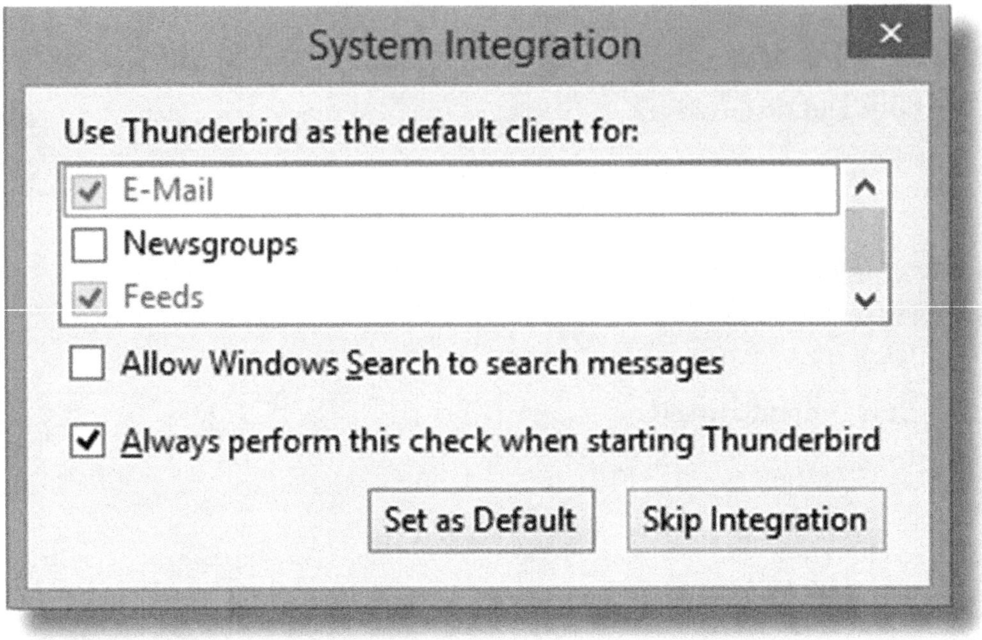

It's up to you. If you're using another email program regularly on this computer, you may want that program to be your default client.

Configuring Thunderbird for a Hotmail account

When you first install Thunderbird, there are no email accounts configured. Thunderbird asks if you'd like to create a new email address with one or more pre-selected providers. Since we already have an account - the Hotmail account we're backing up - we don't need this. Click on **Skip this and use my existing email**.

Thunderbird then presents the Mail Account Setup dialog:

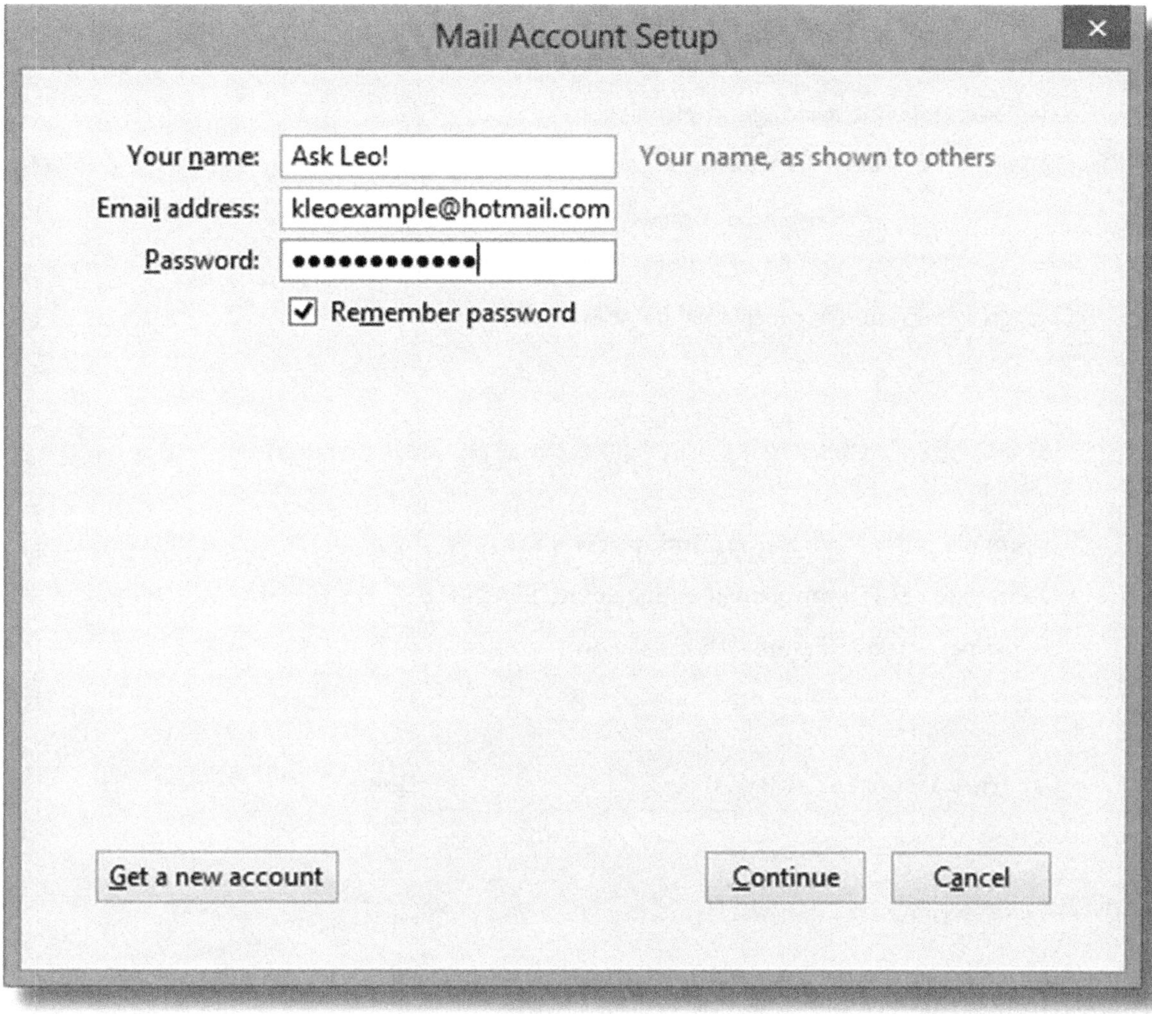

Enter your name, email address, and password. In my case I've entered "Ask Leo!" as my name, "askleoexample@hotmail.com" as my email address, and my password for that account.

Select **Remember password** if this computer is secure and you don't want to type your password every time you fire up Thunderbird.

Click on Continue, and Thunderbird will consult its own database of email providers for the correct settings. Hotmail is in that database.

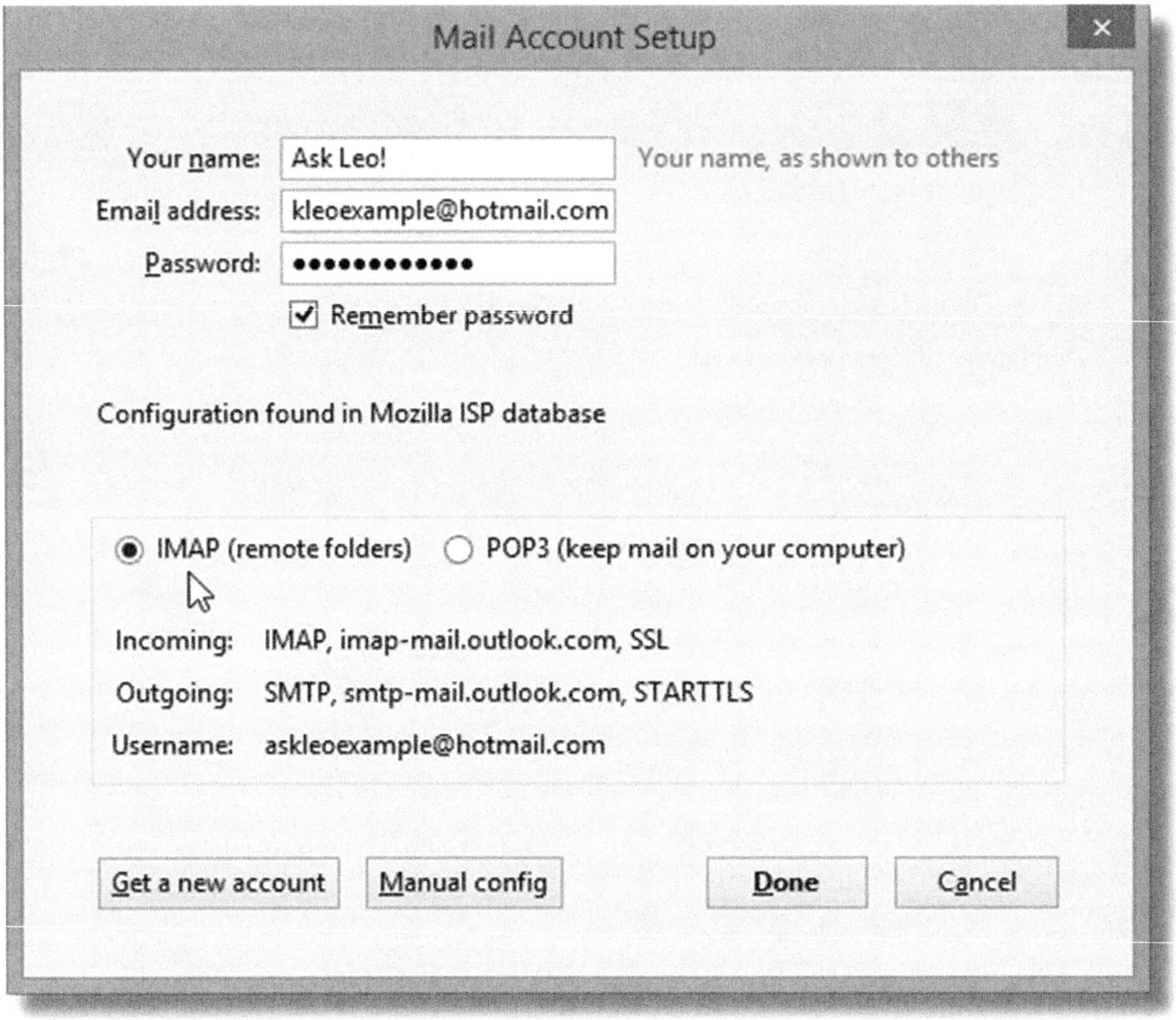

Make sure that IMAP is selected, and click **Done**.

Thunderbird is now configured to access your Hotmail account, and will begin downloading your mail.

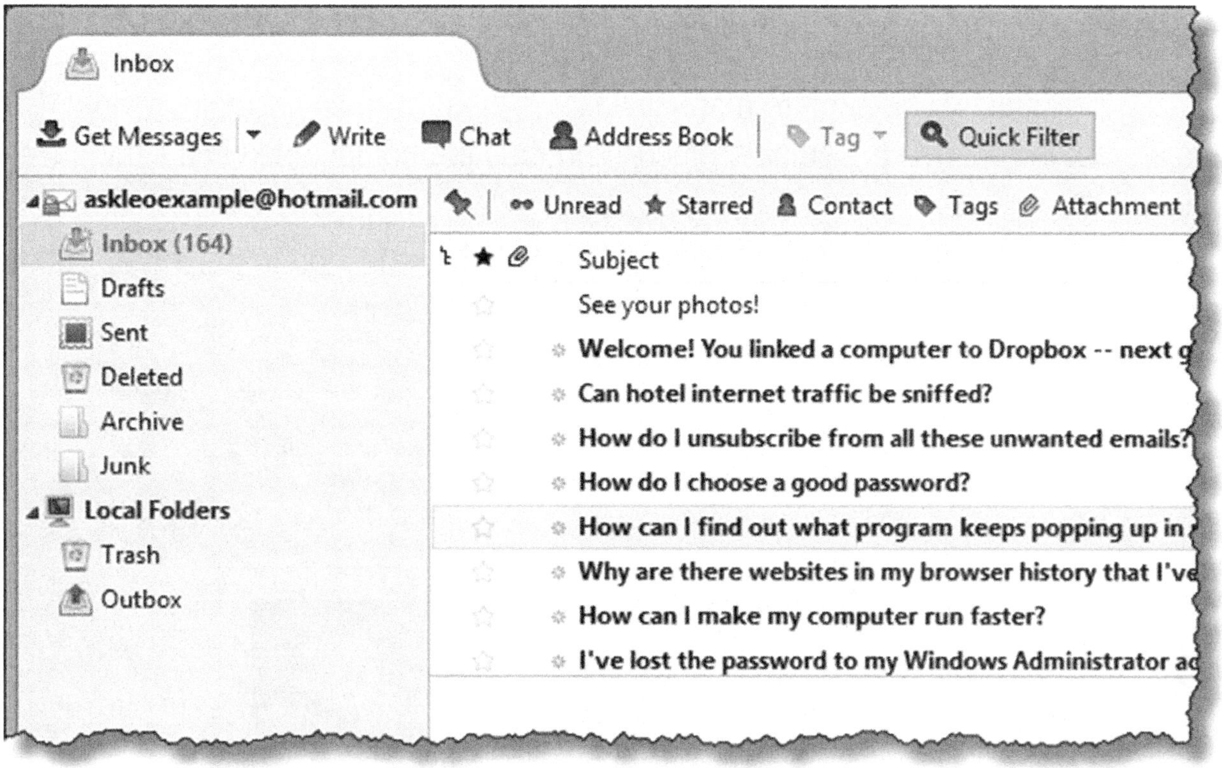

Using Thunderbird

The benefit of using IMAP is that the email program will download a copy to your computer, creating a backup of your email. You can continue to use your email as you normally do via the Outlook.com web interface, your mobile device, or wherever else you've been accessing it.

You can, if you like, use Thunderbird to actually read, compose, and reply to messages. Email you read here will be marked as read elsewhere. In fact, any change you make, such as moving emails to folders, deleting emails, or otherwise managing your email messages will show up in all the places you access your email.

Since IMAP mirrors the activity on your online account down to your PC, note that it will mirror any deletions as well. Delete a message online, and the next time you start Thunderbird, the message will be deleted locally as well. Backing up your PC regularly, as we've already set up, provides you with an additional safety net, since the Thunderbird repository would be included in those backups as well.

Backing up contacts

To back up contacts, you need to manually export them from your email provider. Contacts are not included in the email protocol, and thus are not included in anything we've done so far.

In Outlook.com, contacts are "People". Click on the menu icon to the left of Outlook.com in its header, and then click on the resulting **People** icon.

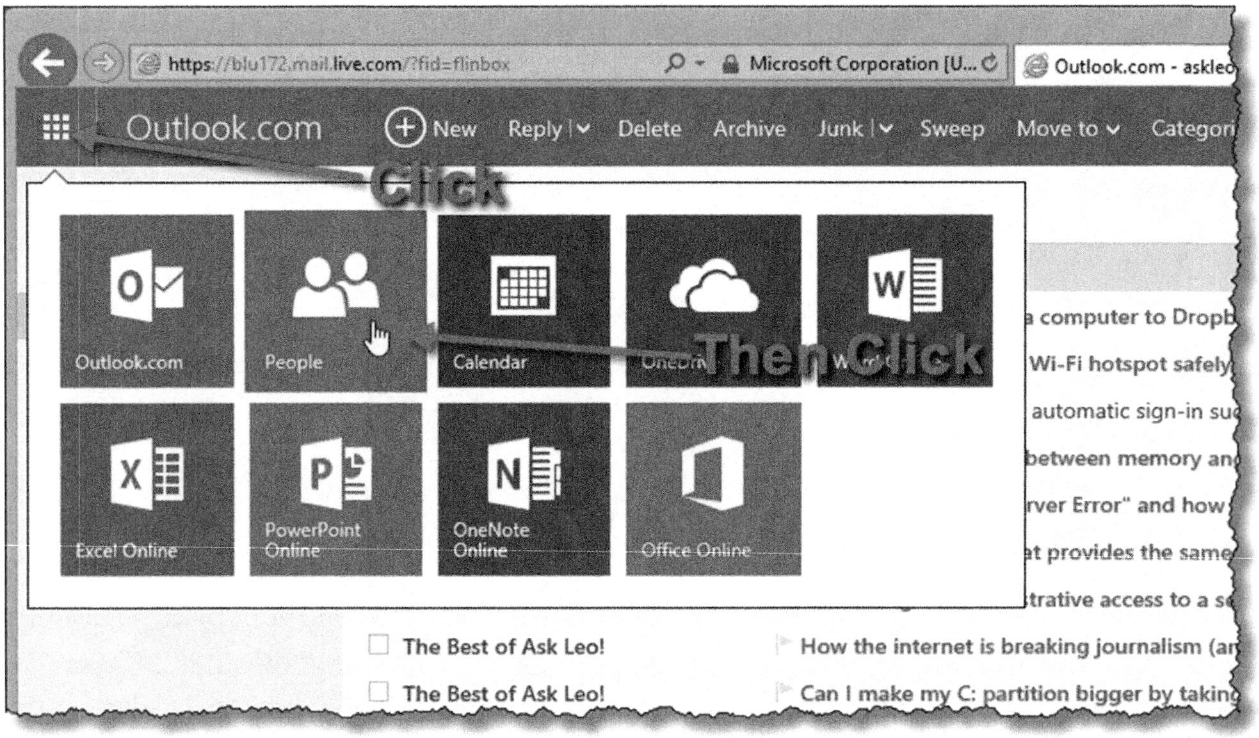

On the People page, click the **Manage** menu item, and then click on **Export for Outlook.com and other services**. Since this is a backup, we choose this format in case we ever need to import the contacts back into Outlook.com.

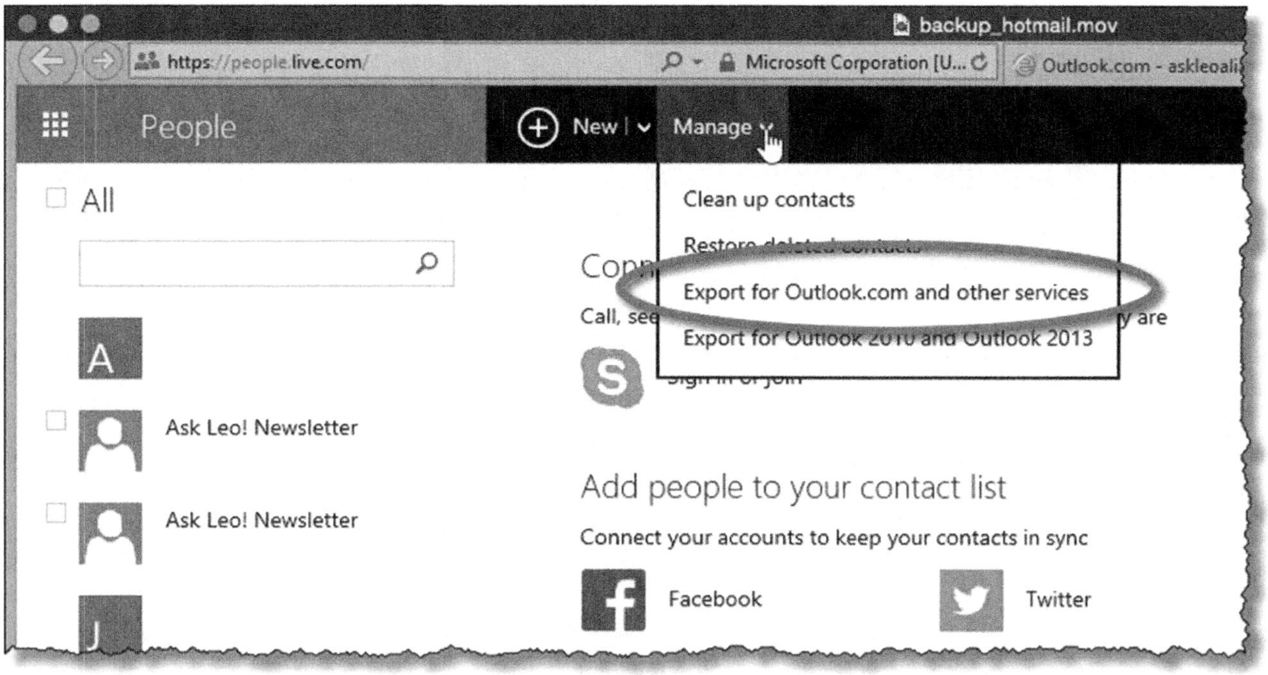

Save the downloaded file somewhere on your computer, ideally in a place that will also be backed up by your regular PC backups.

You can, if you like, import these contacts into Thunderbird, but since our goal here is to back them up, we've achieved that already, simply by downloading the file to our computer[17].

Back up your email

If you don't use Thunderbird regularly, it's important to run Thunderbird periodically to perform this backup. Once a day, once a week, or a couple of times a month, depending on how current you want your backup to be, run Thunderbird, and keep it running until it's downloaded a copy of all updates to your mail.

Similarly, since there's no automated download of your contacts, you'll need to export and save them regularly as well.

17 Importing contacts from one service or program into another, while possible, is notoriously error prone, and almost always results in some amount of lost data. This has to do with the lack of standards about what information is included in a "Contact", and how that information should be represented.

The good news here is that your email is backed up! Particularly when used in conjunction with ongoing PC backups, you're well protected against data loss due to account theft or other problems.

But, in all honesty, I still wouldn't keep the only copy of my master's thesis here. That's worth a few more copies in other locations for safekeeping.

Just Do This

If you use an on-line email account like Hotmail, Outlook.com, Gmail, Yahoo!, or others, install a desktop email program like Thunderbird to automatically download and back up your email using IMAP.

8

Set Yourself Some Reminders

If you've done everything we've talked about so far, congratulations! You're in fantastic shape. Your computer is getting backed up, your work in progress is safe from almost anything, and it would be difficult for you to lose email.

Welcome to the 10% club. In my experience you're better protected than 90% of average computer users today.

But…

Our plan so far does have a couple of rough edges. There are rough edges that I want to make sure you're aware of, and don't overlook. In particular some periodic manual work on your part may be required.

As a result I recommend that you set yourself a reminder to look at the following items once a month or so.

The amount of space left on your external backup drive

You've automated creating periodic backups and those backups use accumulate more and more disk space as they accumulate. What hasn't been done is automating any kind of disk space management.

Once a month see how large the most recent backup image is, and make sure that there's at least 1.5 times that amount of free space left on the external backup drive. If not, then the most common solution is to delete the oldest backup image. *Note*: do not delete the "initial" backup image that we created in chapter 2. That's a backup image you'll want to keep as long as you have this particular computer. Delete instead the oldest "regular" automated backup that you set up.

Another alternative is to move excess backup images to another location or another drive. This typically isn't worth the effort, however. Since you're taking periodic full-image backups you really only need to keep the last couple of months worth to get almost all the protection you could possibly want.

The "automated" alternative is to spring for the paid version of Macrium Reflect. It includes automatic disk space management options that will let you define how many backups to keep. Reflect can automatically delete older backups for you. (Full details and instructions are included in my book Saved! – Backing Up with Macrium Reflect.)[18]

The amount of space you're using in Dropbox

If you're using the default free plan with Dropbox you may eventually find that you've exceeded your two gigabytes of available Dropbox space.

There are several approaches and alternatives.

The most common and practical is simply to move some of the files in your Dropbox folder to other locations on your computer. Files that don't need quite as frequent backing up would be candidates for this kind of archiving.

Naturally you could upgrade to a paid Dropbox account to increase storage space.

You could switch to one of the other services similar to Dropbox that includes more disk space in its free plan. You may have access to another service by virtue of having purchased other software: Microsoft OneDrive comes to mind, as a side effect of purchasing certain versions of Microsoft Office. Just make sure that whatever service you choose has the same file-synchronization tools that will automatically upload files from your PC to their servers as you make changes to those files.

18 http://go.askleo.com/savedmr

Fire up Thunderbird

In reality, you probably want to run Thunderbird more often than once a month, but I include it in this list as a reminder to run it *at least* this often. Make sure to let it run until it's completely synchronizing and downloading all of your email. Depending on just how much email that is that could mean leaving the program running for just a few minutes to several hours.

Given the importance of email to most people, I recommend that you download your email for backup purposes at least once a week.

Back Up Contacts

This is the one portion of the entire process for which there's no good automated solution - certainly no solution that works across all email providers.

Once a month export your contacts (as outlined in the previous chapter).

And, yes, this is something I do myself, both for my email account as well as my wife's. (I use RememberTheMilk.com to set up a monthly reminder. You can even have it email you. Any reminder service or calendar you use will do the trick.)

How Do I Restore?

You've taken the eight steps from the eight preceding chapters to get yourself fully backed up.

I've said it before and I'll say it again: *that's awesome*! If you've followed these steps the chances of losing important data has fallen to near zero.

This book has focussed on backing up… but what happens when you need to restore data?

What happens when something goes wrong?

The answer is: it depends.

It depends on what went wrong.

It depends on what data was potentially lost.

It depends on a lot of different things.

What I *don't* want to do is confuse or leave you feeling overly intimidated by all the different things that can possibly go wrong, and all the different ways you might recover from them.

What's important here is that by following the instructions in this book *you are backed up*. When and if something goes wrong you'll have options - many more options than you had before - to recover that what would otherwise be lost.

Recovery examples

I do, however, want to leave you with some peace of mind. To that end, here are some common disasters, and brief overviews of how you can now recover from them.

Your computer's hard disk dies

- Replace the hard drive.
- Use Macrium Reflect (possibly on another computer) to create a rescue CD.
- Boot your computer from that rescue CD.
- Restore the most recent backup image to your computer.
- Recover your most recent documents from Dropbox.
- Carry on....

Your computer is infected with malware that you can't remove

- Use Macrium Reflect to create a rescue CD.
- Boot your computer from that rescue CD.
- Restore the most recent backup image to your computer.
- Recover your most recent documents from Dropbox.
- Carry on....

You accidentally delete a document you're working on

- Login to Dropbox on the web
- Use Dropbox's file history to recover the deleted document
- Carry on....

Your email account is hacked, and unrecoverable

- Create a new email account.
- Connect Thunderbird to that new account.
- Upload all your email (if you care to) to the new account.
- Restore your contacts from the backups you've been making.
- Carry on...

Your email account is hacked. You're able to get it back, but email and contacts are now missing. What's worse is that you updated your backup in Thunderbird before you realized it, and now everything is missing there as well:

- Restore your Thunderbird folders from the most recent image back up of your computer.
- Upload your email back to your account.
- Restore your contacts from the backups you've been making.
- Carry on…

Yes, when the time comes you may need to reach out for help with some of these scenarios to do them properly, but the bottom line is that you *can* recover *because you're backed up.*

It's possible you may never need any of this, but it's much more likely that at some point, you will.

And you'll be prepared.

And, dare I say it, probably just a little grateful that you are.

Afterword

I hope this book helps you and that you're now part of the 10% club I mentioned - fully backed up and protected from disaster.

If you have found this book valuable, I'd really appreciate it if you posted a review on Amazon.

If you find what you believe to be an error in this book, please register your book (the details are in an upcoming capter) and then visit the errata page for this book. That page will list all known errors and corrections, and give you a place to report anything you've found that isn't already listed.

The Ask Leo! Newsletter

If you found this book helpful, you'll *love* my weekly newsletter.

Each week you'll find fixes to common problems, tips to keep your computer and online information safe and secure, commentary on technology issues of the day, and even the occasional explanation as to just why things are the way they are. It's educational, fun, and can help you be more effective, less frustrated, and safer as you use technology. And it's completely FREE.

Subscribe today and get a FREE bonus copy of my special report: **10 Reasons Your Computer Is Slow (and what to do about it)**

Visit http://go.askleo.com/justbackupnews to learn more, browse the archives, sign up and get your *free* report today!

Register Your Book!

Having purchased this book, you're entitled to additional updates, errata, and other *bonus materials:*

- The companion videos that *show you* the steps discussed in the book.
- Regardless of how you purchased this book, you can download this book in any or all of *three digital formats*: PDF (for your computer or any device that can view PDF files), .mobi (ideal for the Amazon Kindle) or .epub (for a variety of other electronic reading devices).
- Other updates, bonuses and supplementary material I might make available in the future.

Registering gives you access to it all.

Visit http://go.askleo.com/jbureg right now to register.

That link is mentioned only here and it's totally FREE to owners of this book.

NOTE for PDF purchasers:

If you purchased the PDF version of this book directly from the Ask Leo! Store (https://store.askleo.com) *you may already be registered*.

You should have received an email with the information required to sign in. If not, simply use the "Lost your Password?" link at https://members.askleo.com to reset your password using the same email address you used when you purchased the book. If it tells you "There is no user registered with that email address." then simply use the registration link above to create a new account.

Again this applies only if you've purchased the PDF *directly from the Ask Leo! Store.* Purchases made through other outlets need to register using the link above.

About the Author

Why I do what I do

I believe personal technology is key to humanity's future. It has an amazing potential to empower individuals, but it can also frustrate and intimidate.

I want to make technology work for you.

I want to replace that frustration and intimidation with the amazement and wonder that I feel every day.

I want it to be a resource rather than a roadblock; a valuable tool instead of a source of irritation.

I want personal technology to empower you so you can be a part of that amazing future.

I just want it to work, for you.

That's why Ask Leo! exists.

What's in it for you?

I write books, I answer questions, and I educate on important technology topics and trends.

And I just generally share my passion and excitement for the possibilities I believe personal technology holds.

That means you'll find one of the most important and hard-to-find things on the internet: answers, written in a way that you can understand.

I may not have every answer, but I do try to make sure that I take on the important stuff – the questions that people are asking most often and the most important things you can be doing right now to keep your computer – and yourself – empowered and safe.

The important stuff that take you from fear and frustration to "It just works".

I want to make technology work for everyone.

Starting with you.

Why Leo?

Because I love this stuff. I'm one of those weird people that play with computers because I really, really enjoy it. It's both my career and my hobby.

I've been in the personal computer and software industry since 1979, as a software engineer, a manager of software engineers, as a consultant, and as a writer.

In 1983, I joined what was then a medium sized local company called Microsoft, and spent the next 18 years in a wide variety of groups working on a wide variety of software. If you're running Microsoft Windows, have used a Microsoft development tool or Microsoft Money, or purchased a ticket through Expedia, there's a good chance you've been touched by some of my work.

And of course, since 2003, I've been helping to make technology work here on Ask Leo!

If you want to see what else I'm up to, check out my online business card (http://leonotenboom.com), which lists many of my other current roles.

Feedback, Questions and Contacting Leo

I'd love to hear from you.

Honest.

I truly appreciate reader input, comments, feedback, corrections, and opinions – even when the opinions differ from my own!

Here's how best to contact me:

- If you have a comment or a question about this book, I strongly encourage you to register your book, as outlined in a previous chapter, and use the prioritized comment form in the registered owners center.
- If you prefer not to register your book you can email me at leo@askleo.com.
- If you have a computer or tech-related question, the best approach by far is to first search Ask Leo! (https://askleo.com). Many, many questions are already answered right there, and finding those is much faster than waiting for me.
- If you can't find your answer using Search, visit http://askleo.com/book and submit your question. That's a special form just for book purchasers and newsletter subscribers, and it gets prioritized attention.
- If you just want to drop me a line or have something to share that isn't covered above, you can use http://askleo.com/book or email leo@askleo.com for that, too.
- If you're just not sure what to do … email leo@askleo.com.

Copyright & Administrivia

Please note that much of this publication is based on my own personal experience and anecdotal evidence. Although I've made every reasonable attempt to achieve complete accuracy of the content in this book, I can assume no responsibility for errors or omissions. You should use this information as you see fit and at your own risk.

Any trademarks, service marks, product names, or named features are assumed to be the property of their respective owners. They are used only for reference. Unless specifically stated otherwise, there is no implied endorsement if I use one of these terms.

Sharing this Document

The bottom line is that you shouldn't.

More specifically, you shouldn't *make copies and given them to others*.

Loan your copy as you see fit. (Back it up, of course!) However, making an additional copy to *give* to someone else is a no-no. (The rule's pretty simple: if you *loan* the book, they have access to it, and you shouldn't, until they return it. If both you and your friend can use the book at the same time, then you've made a *copy* and that's the part that's wrong.) That goes for uploading a copy to an electronic bulletin board, website, file sharing or similar type of service.

The information in this document is copyrighted. That means that giving copies to others is actually *illegal*. More important than that, it's simply wrong.

Instead, if you think it's valuable enough to share, then encourage your friends who need this book to buy a copy of their own. Or, heck, buy one as a gift for them.

Remember, it's the sale of valuable information in books like this one that makes Ask Leo! possible. It's pretty simple, really; if enough people disregard that, then there'd be no more books, and eventually no more Ask Leo!.

More Ask Leo! Books

The Ask Leo! Guide to Staying Safe on the Internet

The Ask Leo! Guide to Routine Maintenance

Just Do This: Back Up!

Saved! - Backing Up with Macrium Reflect

Saved! - Backing Up with Windows 7 Backup

Saved! - Backing Up with Windows 8 Backup

Backing Up 101 - Six Different Ways to Back Up Your Computer (And Which Ones You Should Use)

Maintaining Windows XP: A Practical Guide

Visit http://askleobooks.com for the latest, up-to-date list!

Printed in Dunstable, United Kingdom

91218874R00065